Contributions of Mexican Mathematicians Abroad in Pure and Applied Mathematics

CONTEMPORARY MATHEMATICS

709

APORTACIONES MATEMÁTICAS
SOCIEDAD MATEMÁTICA MEXICANA

Contributions of Mexican Mathematicians Abroad in Pure and Applied Mathematics

Second Meeting
Matemáticos Mexicanos en el Mundo
December 15–19, 2014
Centro de Investigación en Matemáticas
Guanajuato, Mexico

Fernando Galaz-García
Juan Carlos Pardo Millán
Pedro Solórzano
Editors

AMS AMERICAN MATHEMATICAL SOCIETY

2010 *Mathematics Subject Classification.* Primary 05C50, 58H05, 37H15, 20J06, 37C40, 14H40, 14N10, 30F35, 46E35, 46L54.

Library of Congress Cataloging-in-Publication Data

Names: Galaz-García, Fernando, editor. | Pardo Millán, Juan Carlos, editor. | Solórzano, Pedro, 1981- editor. | Reunión de Matemáticos Mexicanos en el Mundo (2nd : 2014 : Guanajuato, Mexico)

Title: Contributions of Mexican mathematicians abroad in pure and applied mathematics: second meeting, Matemáticos Mexicanos en el Mundo, December 15–19, 2014, Centro de Investigación en Matemáticas, Guanajuato, Mexico / Fernando Galaz-García, Juan Carlos Pardo Millán, Pedro Solórzano, editors.

Other titles: Mexican mathematicians abroad

Description: Providence, Rhode Island: American Mathematical Society, [2018] | Series: Contemporary mathematics; volume 709 | Series: Aportaciones matemáticas / Sociedad Matemática Mexicana | Second meeting of Reunión de Matemáticos Mexicanos en el Mundo, December 15–19, 2014, in Guanajuato, Mexico. | Includes bibliographical references.

Identifiers: LCCN 2017058161 | ISBN 9781470442866 (alk. paper)

Subjects: LCSH: Mathematics–Mexico–Congresses. | Mathematicians–Mexico–Congresses. | Women mathematicians–Mexico–Congresses. | AMS: Combinatorics – Graph theory – Graphs and linear algebra (matrices, eigenvalues, etc.). msc | Dynamical systems and ergodic theory – Random dynamical systems – Multiplicative ergodic theory, Lyapunov exponents. msc | Group theory and generalizations – Connections with homological algebra and category theory – Cohomology of groups. msc | Dynamical systems and ergodic theory – Smooth dynamical systems: general theory – Smooth ergodic theory, invariant measures. msc | Algebraic geometry – Curves – Jacobians, Prym varieties. msc | Algebraic geometry – Projective and enumerative geometry – Enumerative problems (combinatorial problems). msc | Functions of a complex variable – Riemann surfaces – Fuchsian groups and automorphic functions. msc | Functional analysis – Selfadjoint operator algebras (C^*-algebras, von Neumann (W^*-) algebras, etc.) – Free probability and free operator algebras. msc

Classification: LCC QA27.M49 M49 2018 | DDC 510–dc23

LC record available at https://lccn.loc.gov/2017058161

DOI: http://dx.doi.org/10.1090/conm/709

---

10 9 8 7 6 5 4 3 2 1 23 22 21 20 19 18

Contents

Preface

This volume contains the proceedings of the Second Meeting of Mexican Mathematicians Abroad (II Reunión de Matemáticos Mexicanos en el Mundo), which took place at the Centro de Investigación en Matemáticas (CIMAT) in Guanajuato, Mexico, on December 15–19, 2014. This was the second in a series of ongoing biannual meetings aimed at showcasing the research of Mexican mathematicians based outside of Mexico as well as at strengthening academic connections both among the participants and with Mexican institutions. These conferences have been well received by the Mexican mathematical community, and they have had a visible positive impact on many levels. This can be readily measured by different parameters, such as new collaborations, more international grant proposals, organization and participation in specialized events, and, perhaps most importantly, the recruitment of many of the speakers by Mexican academic institutions.

On this occasion, there were 71 attendees from 40 universities in ten different countries, representing a 26% increase in comparison with the first meeting in 2012. Five days of intense activity saw 20 talks by senior and junior researchers, as well as graduate students, working at institutions in the Americas, Europe, and Oceania. A poster session allowed further participants to present their research. The program was rounded off by several roundtable discussions addressing professional opportunities in Mexico and different aspects of working and studying abroad. The organizing committee was composed of four mathematicians based abroad and four based in Mexico: O. Arizmendi (CIMAT, Mexico), N. Bárcenas (Centro de Ciencias Matemáticas, Universidad Nacional Autónoma de México (UNAM), Mexico), F. Galaz-García (Karlsruhe Institute of Technology (KIT), Germany), J. Malagón (University of Toronto, Canada), M. Moreno Rocha (CIMAT, Mexico), J. C. Pardo Millán, (CIMAT, Mexico), R. Ríos-Zertuche (Max Planck Institute for Mathematics, Germany), and P. Solórzano (Universidade Federal de Santa Catarina, Brazil).

On behalf of the organizers, we would like to thank CIMAT, the Sociedad Matemática Mexicana, the Centro de Ciencias Matemáticas at UNAM, and LAISLA Laboratorio Solomon Lefschetz for their financial support. We would also like to thank Norma Cortés and the Departamento de Eventos of CIMAT for their invaluable logistical support. We also thank the speakers and the participants for their commitment to making the meeting a success. Finally, it is a pleasure to thank the authors and referees who contributed to these proceedings as well as the editorial committees and staff of Aportaciones Matemáticas and Contemporary Mathematics for their support in the preparation of this volume. We hope that its publication will continue to foster the interaction among Mexican mathematicians and their peers throughout the world.

Fernando Galaz-García
KIT

Juan Carlos Pardo Millán
CIMAT

Pedro Solórzano
UNAM

Speakers at the Second Meeting of Mexican Mathematicians Abroad, CIMAT, December 15–19, 2014. (Photo by CIMAT, A.C.)

Organizers at the Second Meeting of Mexican Mathematicians Abroad, CIMAT, December 15–19, 2014. (Photo by CIMAT, A.C.)

Participants at the Second Meeting of Mexican Mathematicians Abroad,
CIMAT, December 15–19, 2014. (Photo by CIMAT, A.C.)

List of Talks and Speakers

- Thom spectra from the ∞-categorical point of view
  Omar Antolín Camarena, Harvard University

- Primitive models of social networks: random trees
  Laura Eslava, McGill University

- Stochastic modeling of carcinogenic tumors via Moran models
  Nicolás Flores Castillo, Rice University

- Equivariant techniques in quantum cohomology
  Eduardo González, University of Massachusetts, Boston

- Evolution: At the boundary between biology and mathematics
  Adrián González Casanova Soberón, Technische Universität Berlin

- Non-autonomous dynamical systems, multiplicative ergodic theorems, and applications
  Cecilia González-Tokman, University of New South Wales

- Methodology for the statistical analysis of data from complex surveys
  Dagmar M. Hajducek, University of Waterloo

- Combinatorial rigidity of the modular Teichmüller group
  Jesús Hernández Hernández, Aix-Marseille Université

- Stability phenomena in cohomology
  Rita Jiménez Rolland, Northeastern University

- Birational geometry in moduli spaces
  César Lozano Huerta, Harvard University

- Kähler geometry in moduli spaces as an extremum in conformal field theory
  Claudio Meneses, Max-Planck-Institut für Mathematik, Bonn

- Describing abelian varieties of dimension 6
  Ángela Ortega, Humboldt-Universität zu Berlin

- A ramification rule via Littlemann paths
  Jacinta Pérez Gavilán, Universität zu Köln

- Travelling waves in a family of degenerate non-linear reaction-diffusion advection equations
  Judith Pérez Velázquez, Helmholtz Zentrum München

- Flows in manifolds of dimension three without periodic orbits
  Ana Rechtman, Université de Strasbourg

- Haris-Chandra bimodules over Cherednik algebras
  José Simental Rodríguez, Northeastern University

- Approximations of homomorphism functions
  Pablo Soberón Bravo, University of Michigan

- Characterizations of measures in the dual of BV
  Mónica Torres, Purdue University

- Rado's conjecture, reflection and semi-stationary reflection principles
  Víctor Torres-Pérez, Technische Universität Wien

- The combinatorics of the circular element
  Carlos Vargas, Universität des Saarlandes

List of Participants

Dayron Fabian Achury Calderón
Fundación Universitaria Konrad Lorenz, Colombia

Omar Antolín Camarena
Harvard University, USA

Octavio Arizmendi
CIMAT, Mexico

Noé Bárcenas Torres
Centro de Ciencias Matemáticas, UNAM, Mexico

Luis Miguel Barrón García
Universidad de Guadalajara, Mexico

Gabriel Hernán Berzunza Ojeda
Universität Zürich, Switzerland

Armando José Cabrera Pacheco
University of Miami, USA

Ivonne Callejas Arellano
Georg-August-Universität Göttingen, Germany

Victoria Cantoral Farfán
Institut de Mathématiques de Jussieu - Université Paris Diderot, France

Hugo Alfredo Carrillo Serrano
CIMAT, Mexico

Luz Angélica Caudillo Mata
University of British Columbia, Canada

Diego Corro Tapia
Instituto de Matemáticas, UNAM, Mexico

Manuel Esteban Enríquez Villeda
UNAM, Mexico

Laura Eslava
McGill University, Canada

Nicolás Flores Castillo
Rice University, USA

José Salvador Flores Hernández
CIMAT, Mexico

Carmen Galaz-García
Universidad de Guanajuato, Mexico

Fernando Galaz-García
Karlsruher Institut für Technologie, Germany

Erik Alejandro Gallegos Baños
Universidad de Guanajuato, Mexico

Tania Gómez Fernández
UNAM, Mexico

Adrián González Casanova Soberón
Technische Universität Berlin, Germany

Eduardo Adolfo González Méndez
University of Massachusetts, Boston, USA

Cecilia González-Tokman
University of New South Wales, Australia

José Ramón Guardiola Espinosa
UNAM, Mexico

Dagmar Mariaca Hajducek
University of Waterloo, Canada

Gerardo Hernández Dueñas
Instituto de Matemáticas, UNAM, Mexico

Jesús Hernández Hernández
Aix-Marseille Université, France

Eduardo Hernández Huerta
Escuela Superior de Física y Matemáticas, IPN, Mexico

Rita Jiménez Rolland
Northeastern University, USA

Bryan David Juárez Hernández
Facultad de Ciencias, UNAM, Mexico

Oliver Antonio Juárez-Romero
CIMAT, Mexico

Michelle Lagunas García
Facultad de Ciencias, UNAM, Mexico

César Lozano Huerta
Harvard University, USA

Francisco Javier Martínez Deferia
CIMAT, Mexico

Guillermo Elías Martínez Dibene
CIMAT, Mexico

Claudio Meneses
Max Planck Institut für Mathematik, Germany

Mariana Míreles González
FES Acatlán, UNAM, Mexico

Jesús Ernesto Montes Rosas
Universidad Autónoma de Ciudad Juárez, Mexico

Mónica Moreno Rocha
CIMAT, Mexico

Hugo Ariel Nava Sauceda
Universidad Autónoma de San Luis Potosí, Mexico

Jesús Núñez Zimbrón
Facultad de Ciencias, UNAM, Mexico

Ángela Ortega
Humboldt-Universität zu Berlin, Germany

Juan Carlos Pardo Millán
CIMAT, Mexico

Hugo Peña Gómez
CIMAT, Mexico

Raquel Perales
SUNY-Stony Brook, USA

Jacinta Pérez Gavilán
Universität zu Köln, Germany

Judith Pérez Velázquez
Helmholtz Zentrum München, Germany

Saraí Ramírez Gómez
FES Acatlán, UNAM, Mexico

Alexandre Ramos Peon
Universität Bern, Switzerland

Ana Rechtman
Université de Strasbourg, France

Araceli Reyes Morales
Instituto de Matemáticas, UNAM, Mexico

Pedro Reyes Pérez
Universidad Autónoma Metropolitana, Mexico

Mohammad Reza Rahmati
CIMAT, Mexico

José Augusto Rodríguez Martínez
Universidad de Guanajuato, Mexico

Julio César Rubén Romo Cruz
IF-UNAM and UACM, Mexico

Mauricio Salinas Rodríguez
Facultad de Ciencias, UNAM, Mexico

Salvador Sierra Murillo
Freie Universität Berlin, Germany

José Simental Rodríguez
Northeastern University, USA

Pablo Soberón Bravo
University of Michigan, USA

Marlem Solís Santana
Universidad Autónoma de Zacatecas, Mexico

Pedro Solórzano
Universidade Federal de Santa Catarina, Brazil

Mónica Torres
Purdue University, USA

Alejandra Torres Manotas
Fundación Universitaria Konrad Lorenz, Colombia

Jonatán Torres Orozco Román
CIMAT, Mexico

Víctor Torres-Pérez
Vienna University of Technology, Austria

Juan Fernando Valdés Cruz
CIMAT, Mexico

Carlos Vargas Obieta
Universität des Saarlandes, Germany

Alejandra Mérida Vázquez
Facultad de Ciencias, UNAM, Mexico

Fernando Velasco Luna
Benemérita Universidad Autónoma de Puebla, Mexico

Arturo Zamudio Alpízar
CIMAT, Mexico

Ilse Beatriz Zubieta Martínez
UAM-Cuajimalpa, Mexico

Contemporary Mathematics
Volume **709**, 2018
http://dx.doi.org/10.1090/conm/709/14288

On bounds for the energy of graphs and digraphs

Octavio Arizmendi and Oliver Juarez-Romero

Abstract. In this note we review the concept of energy of a graph of Gutman and present the mayor results in bounding this energy. We also recall inequalities for the energies of digraphs and matrices. We present relations between these energies and some shorter proofs for some of the known results from the point of view of Non Commutative Probability. Finally, we introduce the concept of energy of a vertex and derive new inequalities.

Contents

1. Introduction

The energy of a graph introduced by Gutman [**13, 14**] is defined as the sum of the absolute values of the eigenvalues of the adjacency matrix. This quantity has applications to chemistry. Certain quantities of importance to chemists, such as the heat of formation of a conjugated hydrocarbon, are related to the π-electron energy that can be approximated by the energy of an appropriate "molecular" graph. Firstly, this quantity was of interest because the calculation of the HMO total π-electron energy was a serious problem and thus giving approximation of this energy was of great help. Later the spirit was to understand how the structural details of a graph influence this quantity.

2010 *Mathematics Subject Classification.* Primary 05C50.

Key words and phrases. Graph energy, non commutative probability, graphs spectra, inequalities.

The authors were supported by CONACYT Grant 222668.

More recently, motivated by Coulson's integral formula [**7**], Peña and Rada [**26**] generalized the concept of energy of a graph to include directed graphs (or digraphs). It turns out that, from the mathematical point of view, studying the energy for digraphs is more interesting. As an example, the absence of cycles in a digraph is characterized by having energy equal to 0.

On the other hand, Nikoforov's observation [**24**] that the energy of a graph is nothing else than the nuclear (or trace) norm of the adjacency matrix naturally leads to a generalization of energy to matrices. This gives a broader vision of the concept of energy. However, in general, these two generalizations do not coincide and then it is interesting to understand how they are related with each other.

Non Commutative Probability is a new area of research which has relations with many other areas of mathematics, such as Operator Theory, Probability, Combinatorics, Representation Theory, among others. In particular, Voiculescu's Free Probability (see [**31**]) has been used to understand the asymptotic behavior of Large Random Matrices [**30**], while Quantum Probability helps to understand Spectral Theory of Graphs (see Hora and Obata [**19**]). However, until now, the energy of a graph has not been studied from the point of view of this theory.

In this note we present a brief summary on the main inequalities given for the energy of graphs, digraphs and matrices and review the concept of energy of a graph of Gutman from the point of view of Non Commutative Probability, which naturally generalize the ideas of Nikiforov. We clarify some of the relations between the energy of a graph, the energy of a digraph and the energy of a matrix in a rather precise way. Finally, by introducing the concept of *energy of a vertex* we give new inequalities for the energy of graph.

The note is organized as follows. In Section 2 we remind basic facts about the graph energy of Gutman. In Section 3 we give preliminaries on Non Commutative Probability. In Section 4 we review known bounds for the energy of a graph of Gutman and we apply the ideas of Non Commutative Probability to obtain new shorter proofs of some of these bounds. Section 5 introduces the concepts of energies for digraphs and matrices. Finally, in Section 6 and 7 we present new results. Section 6 is concerned with relations between energies for graphs, digraphs and matrices. In Section 7 we introduce the notion of *energy of a vertex* and give an improvement of McClelland inequality.

Acknowledgment. We would like to thank the anonymous referee for the careful reading of the paper, which helped much in improving its presentation.

2. Energy of a graph

In this section we present the definition of energy of a graph, its basic properties and describe briefly the chemical motivation of Gutman.

2.1. Energy of a graph. Let G be a graph without loops and multiple edges. The number of vertices and edges of G are denoted by n and m, respectively. Its vertices are labeled by $v_1, v_2, \dots, v_n$. The adjacency matrix $A = A(G)$ of G is a square matrix of order n whose (i, j)-entry is defined as

$$A_{ij} = \begin{cases} 1, & \text{if the vertices } v_i \text{ and } v_j \text{ are adjacent,} \\ 0, & \text{otherwise.} \end{cases} \tag{2.1}$$

The eigenvalues of $A(G)$ are said the eigenvalues of the graph G. A graph on n vertices has n eigenvalues (not all of which need to be distinct); these will be denoted by $\lambda_1, \lambda_2, \ldots, \lambda_n$ and labeled in a decreasing manner: $\lambda_1 \geq \lambda_2 \geq \cdots \geq \lambda_n$. The set of the all n eigenvalues of G forms the spectrum of G.

Since $A(G)$ is selfadjoint then the eigenvalues of the graph G are necessarily real-valued. Moreover, since G has no loops then the trace of A equals 0, i.e., $Tr(A) = \lambda_1 + \cdots + \lambda_n = 0$. Thus, if $\lambda_1^+, ..., \lambda_k^+$ and $\lambda_{k+1}^-, ..., \lambda_n^-$ denote the positive and negative eigenvalues, respectively, then

$$\sum_{j=1}^{k} \lambda_j^+ = - \sum_{j=k+1}^{n} \lambda_j^- . \tag{2.2}$$

Spectral properties of graphs (including properties of the characteristic polynomial) have been extensively studied; for a detailed survey see [**4**].

The concept of graph energy was first defined by Gutman in [**14**] and it emerged from the idea of Hückel energy in theoretical chemistry as we briefly explain below in Section 2.2. For a graph G, the energy of G, denoted by $\mathcal{E}(G)$ is defined as the sum of the absolute values of its eigenvalues,

$$\mathcal{E}(G) = \sum_{i=1}^{n} |\lambda_i|. \tag{2.3}$$

The characteristic polynomial of $A(G)$ is said to be the characteristic polynomial of the graph G. Thus, the characteristic polynomial of G is a monic polynomial of degree n and is defined as

$$\phi(G, x) = \det[xI - A(G)], \tag{2.4}$$

where I stands for the unit matrix of order n. The eigenvalues of the graph are just the zeros of the characteristic polynomial. Coulson [**7**] provided the following integral formula for the energy of a graph, $\mathcal{E}(G)$:

$$\mathcal{E}(G) = \frac{1}{\pi} \int_{-\infty}^{\infty} \left(n - \frac{ix\phi'(G, ix)}{\phi(G, ix)} \right) dx, \tag{2.5}$$

where $\phi(G, x)$ is the characteristic polynomial of graph G, and $\phi'(G, x)$ its derivative.

2.2. Chemical Motivation. We now present the chemistry theory behind the definition of graph energy. We follow the exposition of [**22**]. In the 30's, Hückel suggested a method for approximating the solution to the Schrödinger's equation for a certain class of molecules known as "Hückel molecular orbital (HMO) theory". Recall that Schrödinger's equation is a second-order partial differential equation of the form

$$\hat{H}\Xi = \mathcal{E}\Xi, \tag{2.6}$$

where Ξ is the so-called wave function of the system, $\hat{H}$ the Hamiltonian operator, and $\mathcal{E}$ the energy of the system. When applied to a molecule, the Schrödinger equation enables to describe the behavior of the electrons in this molecule and to establish their energies.

Now, instead of trying to solve (2.6) one tries to approximate $\hat{H}$ by a finite dimensional operator $\mathbf{H}$ and solve the problem

$$\mathbf{H}\Xi = \mathcal{E}\Xi. \tag{2.7}$$

Here $\mathbf{H}$ is matrix with entries defined as follows

$$\mathbf{H}_{ij} = \begin{cases} \alpha, & \text{if } i = j, \\ \beta, & \text{if the atoms } i \text{ and } j \text{ are chemically bonded,} \\ 0, & \text{otherwise.} \end{cases} \tag{2.8}$$

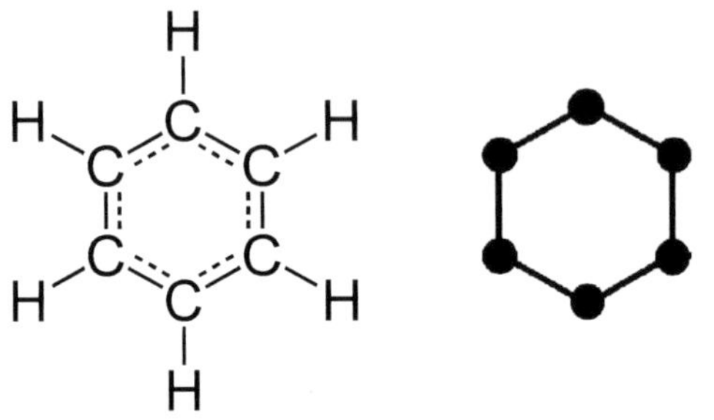

FIGURE 1. A molecule of benzene which has a system of 6 π-electrons together with its associated graph.

It is clear that if one associates to a molecule the graph where $i \sim j$ if i and j are chemically bonded (see Fig. 1), then one has

$$\mathbf{H} = \alpha I + \beta A(G). \tag{2.9}$$

It turns out that, in the HMO approximation, the total energy of all π-electrons is given by

$$\mathcal{E}_\pi = \alpha_n + \beta \sum_{j=1}^{n} g_j \lambda_j, \tag{2.10}$$

where g_j is the so-called occupation number. Since α and β are assumed to be constant, one is interested only in the quantity

$$\mathcal{E} = \sum_{j=1}^{n} g_j \lambda_j. \tag{2.11}$$

By a general physical law, g_j can only assume the values $0, 1$, or 2. However, for most of the chemically relevant cases, one has

$$g_j = \begin{cases} 2, & \text{whenever } \lambda_j > 0, \\ 0, & \text{whenever } \lambda_j \leq 0, \end{cases} \tag{2.12}$$

and thus we arrive at the formula

$$\mathcal{E} = \sum_{j=1}^{k} 2\lambda_j^+ = \sum_{j=1}^{n} |\lambda_j|. \tag{2.13}$$

3. Non commutative probability

A *non commutative probability space* is a pair $(\mathcal{A}, \tau)$, where $\mathcal{A}$ is an algebra with unity $1_{\mathcal{A}}$ and τ is a linear functional such that $\tau(1_{\mathcal{A}}) = 1$.

The elements of $\mathcal{A}$ are called (non-commutative) random variables. An element $a \in \mathcal{A}$ such that $a = a^*$ is called *self-adjoint.* If τ satisfies that $\tau(ab) = \tau(ba)$ for all a, b in $\mathcal{A}$ we say that τ is *tracial* and call it a trace. Finally, we say that τ is *positive* if $\tau(a^*a) \geq 0$.

The functional τ should be understood as the expectation in classical probability.

If $\mathcal{A}$ is a unital C^*-algebra and $\tau : \mathcal{A} \to \mathbb{C}$ is a positive unital linear functional then the pair $(\mathcal{A}, \tau)$ is called a *C^*-probability space.* From now on we will assume that the non commutative probability spaces considered here are of this type.

For $a \in \mathcal{A}$, we will refer to the value of $\tau(a^k)$, $k \in \mathbb{N}$, as the k-th moment of a. For any self-adjoint element $a \in \mathcal{A}$, by the Riesz-Markov-Kakutani representation theorem, there exists a unique probability measure μ_a (its spectral distribution) with the same moments as a, that is,

$$\int_{\mathbb{R}} x^k \mu_a(dx) = \varphi(a^k), \quad \forall k \in \mathbb{N}. \tag{3.1}$$

Let $p > 0$, the L_p norm of a the random variable $a \in A$ is given by

$$||a||_p = \tau(|a|^p)^{1/p},$$

where $|a| = (aa^*)^{1/2}$ denotes the absolute value of a. The case $p = 1$ is called the *trace norm* or nuclear norm and will be denoted by $||a||_*$. It satisfies $||a||_* = \tau(|a|) = \tau(|a^*|) = ||a^*||_*$.

Let $0 < r, p, q \leq \infty$ be such that $1/r = 1/p + 1/q$. The usual Hölder inequality extends to the non-commutative setting in the following sense.

$$||ab||_r \leq ||a||_p ||b||_q. \tag{3.2}$$

In particular for $p = 2$ and $q = 2$, since $\tau(|ab|) \geq |\tau(ab)|$ we get the Cauchy-Schwarz inequality,

$$|\tau(ab)| \leq \tau(aa^*)^{1/2} \tau(bb^*)^{1/2}. \tag{3.3}$$

Also taking $b = 1$ we get for $p \leq r$

$$||a||_r \leq ||a||_p. \tag{3.4}$$

In this note we will only consider two types of C^*-probability spaces. The first one is canonical, the non commutative probability space $(\mathcal{M}_n, tr)$, where $\mathcal{M}_n$ is the set of matrices of size $n \times n$ and, for a matrix $M \in \mathcal{M}_n$, tr denotes the normalized trace and is given by

$$tr(M) = \frac{1}{n} Tr(M) = \frac{1}{n} \sum_{i=1}^{n} M_{ii}. \tag{3.5}$$

The second is the space $(\mathcal{M}_n, \phi_i)$, where $\mathcal{M}_n$ is the set of matrices of size $n \times n$ and, for a matrix $M \in \mathcal{M}_n$, ϕ_i evaluated in M is given by

$$\phi_i(M) = M_{ii}. \tag{3.6}$$

Both of these linear functionals have nice relations when applied to graphs. Let $G = (V, E)$ be a finite graph with vertex set $\{1, \ldots, n\}$ and let $A(G)$ be the adjacency matrix. We denote by $\mathcal{A}(G) \subset \mathcal{M}_n$ the adjacency algebra, i.e., the $*$-algebra generated by $A(G)$.

It is easy to see that the k-th moment of A with respect to ϕ_i is given by the number of walks in G of size k starting and ending at the vertex i. That is,

$$\phi_i(A^k) = |\{(w_1, ..., w_k) : w_1 = w_k = v_i \text{ and } (w_j, w_{j+1}) \in E, \ \forall j \in [k-1]\}|. \tag{3.7}$$

Similarly, since $tr(A) = \frac{1}{n}\sum_{i=1}^{n} \phi_i(A)$ then the k-th moment of A with respect to tr gives the expected number of closed walks of a point chosen uniformly at random. That is

$$Tr(A^k) = |\{(w_1, ..., w_k) : w_1 = w_k \text{ and } (w_j, w_{j+1}) \in E, \ \forall j \in [k-1]\}|. \tag{3.8}$$

Thus one can get combinatorial information of G from the values of ϕ_1 in elements of $A(G)$ and vice versa.

By multiplying by n in both sides of Hölder inequality for the normalized trace we can obtain the known Hölder inequality for the trace of a matrix,

$$Tr(|MN|) \leq Tr(|M|^p)^{1/p} Tr(|N|^q)^{1/q}. \tag{3.9}$$

Finally, note that for a graph with n vertices we have that

$$\mathcal{E}(G) = ntr(|A(G)|) = \sum_{i=1}^{n} \phi_i(|A(G)|). \tag{3.10}$$

4. Bounds for the energy of graphs

In this section, we focus on some of the most important bounds for the energy of graphs. We present some of the original proofs of these upper bounds.

4.1. First inequalities.

One of the first inequalities was given by McClelland in [**23**]. He considered the n vertices and m edges of a graph G for the following upper bound.

THEOREM 4.1 ([**23**]). *For any graph G with n vertices and m edges,*

$$\mathcal{E}(G) \leq \sqrt{2mn}. \tag{4.1}$$

PROOF. Recall the Cauchy-Schwarz inequality

$$\left(\sum_{i=1}^{N} a_i b_i\right)^2 \leq \left(\sum_{i=1}^{N} a_i^2\right)\left(\sum_{i=1}^{N} b_i^2\right), \tag{4.2}$$

which holds for arbitrary real-valued numbers a_i, b_i and $i = 1, 2, \ldots, N$.

Thus, choosing in (4.2) $N = n$, $a_i = |\lambda_i|$, $b_i = 1$ and using that $\sum_{i=1}^{n} \lambda_i^2 = 2m$, we get the desired result.

□

ALTERNATIVE PROOF. If we apply (3.4) for $r = 1$ and $p = 2$ we get that

$$\frac{E(G)}{n} = tr(|A|) \leq tr(|A|^2)^{1/2} = (\frac{2m}{n})^{1/2}. \tag{4.3}$$

□

Later, Koolen and Moulton [**20**] improved the upper bound of McClelland by a more clever use of (4.2).

THEOREM 4.2 ([**20**]). *For any graph G with n vertices and m edges. If $2m \geq n$, then*

$$\mathcal{E}(G) \leq \frac{2m}{n} + \sqrt{(n-1)\left[2m - \left(\frac{2m}{n}\right)^2\right]}. \tag{4.4}$$

Equality in (4.4) holds if $G \cong \frac{n}{2}K_2$, $G \cong K_n$ or G is a non-complete connected strongly regular graph[1] *with two eigenvalues both having absolute values equal to* $\sqrt{\frac{2m-\left(\frac{2m}{n}\right)^2}{n-1}}$.

PROOF. Choosing $N = n-1$, $a_i = |\lambda_{i+1}|$, $b_i = 1$ in (4.2), we have that

$$\left(\mathcal{E}(G) - \lambda_1\right)^2 \leq \left(2m - \lambda_1^2\right)(n-1), \tag{4.5}$$

which, taking square roots and rearranging, yields

$$\mathcal{E}(G) \leq \lambda_1 + \sqrt{(n-1)\left(2m - \lambda_1^2\right)}. \tag{4.6}$$

Next, we consider the function $f(x) = x + \sqrt{(n-1)\left(2m - x^2\right)}$. By elementary calculus we can see that f is monotonically decreasing in the interval $(\sqrt{2m/n}, \sqrt{2m})$. If $\frac{2m}{n} \geq 1$, then $\frac{2m}{n} \leq \lambda_1$ (see [**6**]). Thus,

$$\mathcal{E}(G) \leq \frac{2m}{n} + \sqrt{(n-1)\left[2m - \left(\frac{2m}{n}\right)^2\right]}. \tag{4.7}$$

□

By maximizing (4.4) with respect to m one arrives to the following corollary.

COROLLARY 4.3. *([**21**]) Let G be a graph on $n > 1$ vertices. Then*

$$\mathcal{E}(G) \leq \frac{n}{2}\left(\sqrt{n} + 1\right). \tag{4.8}$$

Equality in (4.8) holds if and only if G is a strongly regular graphs with parameters $\left(n, \frac{n+\sqrt{n}}{2}, \frac{n+2\sqrt{n}}{4}, \frac{n+2\sqrt{n}}{4}\right)$.

Another inequality in terms of the vertex degree sequence $d_1, d_2, \ldots, d_n$ of G that follows these ideas and the fact that $\lambda_1 \geq \sqrt{\frac{\sum_{i=1}^n d_i^2}{n}}$ was found in [**32**],

$$\mathcal{E}(G) \leq \sqrt{\frac{\sum_{i=1}^n d_i^2}{n}} + \sqrt{(n-1)\left[2m - \left(\frac{\sum_{i=1}^n d_i^2}{n}\right)\right]}. \tag{4.9}$$

Equality in (4.9) holds if and only if G is either $\frac{n}{2}K_2$ (if $n = 2m$), K_n (if $m = n(n-1)/2$), a noncomplete connected strongly regular graph with two nontrivial eigenvalues both with absolute value $\sqrt{\left[2m - \left(\frac{2m}{n}\right)^2\right]/(n-1)}$ or nK_1 (if $m = 0$).

[1] A strongly regular graph with parameters (n, d, λ, μ) is a d-regular graph with n vertices such that every two adjacent vertices have λ common neighbors and every two non-adjacent vertices have μ common neighbors.

4.2. The energy of bipartite graphs. When restricted to certain kind of graphs one may improve the upper bounds by considering the details in the structure of the graph. In particular, Koolen and Moulton [**21**] gave the following inequality for bipartite graphs. The main observation is that in this case since $\lambda_1 = \lambda_n$ then one can take $N-2$ instead of $N-1$ in (4.9), as in the proof of Theorem 4.2.

THEOREM 4.4 ([**21**]). *If $2m \geq n$ and G is a bipartite graph with $n > 2$ vertices and m edges, then*

$$\mathcal{E}(G) \leq 2\left(\frac{2m}{n}\right) + \sqrt{(n-2)\left[2m - 2\left(\frac{2m}{n}\right)^2\right]}. \tag{4.10}$$

Equality holds in (4.10) if and only if at least one of the following statements holds:

(i) *$n = 2m$ and $G = mK_2$;*
(ii) *$n = 2t$, $m = t^2$, and $G = K_{t,t}$;*
(iii) *$n = 2v$, $2\sqrt{m} < n < 2m$, and G is the incidence graph of a symmetric $2-(v,k,\lambda)$-design*[2] *with $k = \frac{2m}{n}$ and $\lambda = \frac{k(k-1)}{v-1}$.*

By maximizing (4.10) with respect to m one arrives to the following corollary.

COROLLARY 4.5. *([**21**]) Let G be a bipartite graph on $n > 2$ vertices. Then*

$$\mathcal{E}(G) \leq \frac{n}{\sqrt{8}}\left(\sqrt{n} + \sqrt{2}\right). \tag{4.11}$$

Equality holds in (4.11) if and only if $n = 2v$ and G is the incidence graph of a $2-(v, \frac{v+\sqrt{v}}{2}, \frac{v+2\sqrt{v}}{4})$-design.

A lower bound in terms of the rank of the adjacency matrix of a bipartite graph is given by Akbari et al. [**1**].

THEOREM 4.6 ([**1**]). *If G is a connected bipartite graph of rank r, then*

$$\mathcal{E}(G) \geq \sqrt{(r+1)^2 - 5}. \tag{4.12}$$

4.3. Spectral Moments. The spectral moments method is an important technique that permits to give lower bounds for the energy of a graph. The first applications of this method can be found in [**9, 10, 32**]. We define for a graph with n vertices the moment-like quantities

$$M_k^* = M_k^*(G) = \sum_{i=1}^{n} |\lambda_i|^k. \tag{4.13}$$

The following theorem was reported by Rada and Tineo [**29**] for bipartite graphs and for all graphs by de la Peña, Mendoza and Rada [**10**]. We present a new shorter proof of this theorem.

[2] A $2-(v,k,\lambda)$-design is a collection of k-subsets called blocks of a set of v points, such that each 2-set of points lies in exactly k blocks. The design is called symmetric if the number of blocks equals v. The incidence matrix B of a $2-(v,k,\lambda)$-design is the $v \times b$ matrix defined by setting, for each x a point and S a block, $B_{x,S} = 1$ if $x \in S$ and 0 otherwise. The incidence graph of a design is defined to be the graph with adjacency matrix $\begin{pmatrix} 0 & B \\ B^T & 0 \end{pmatrix}$.

THEOREM 4.7. *Let G be a graph of order n with at least one edge. Then*

$$\mathcal{E}(G) \geq \frac{(M_2^*(G))^2}{\sqrt{M_2^*(G) M_4^*(G)}}. \tag{4.14}$$

PROOF. From (3.9) we see that taking $p = 3$ and $q = 3/2$, for $M = |A|^{4/3}$ and $N = |A|^{2/3}$, we obtain

$$Tr(|A|^2) \leq Tr(|A|)^{2/3} Tr(|A|^4)^{1/3}, \tag{4.15}$$

from where

$$\mathcal{E}(G) = Tr(|A|) \geq Tr(|A|^2)^{3/2} Tr(|A|^4)^{-1/2}. \tag{4.16}$$

□

From the proof of the previous result one easily realizes the following.

THEOREM 4.8. *Let G be a graph of order n with at least one edge. Then*

$$\mathcal{E}(G) \geq \frac{(M_u^*(G))^p}{(M_{q(u-1)+1}^*(G))^{p/q}}. \tag{4.17}$$

A rather sophisticated generalization of (4.14) by Zhou et al. [**33**] which follows from the last theorem is the following.

THEOREM 4.9 ([**33**]). *Let G be a graph of order n with at least one edge, and let r, s, t, be nonnegative real numbers, such that $4r = s + t + 2$. Then*

$$\mathcal{E}(G) \geq M_r^*(G)^2 [M_s^*(G) M_t^*(G)]^{-\frac{1}{2}}. \tag{4.18}$$

Equality holds in (4.18) if and only if the components of the graph G are isolated vertices and/or complete bipartite graphs $K_{p_1,q_1}, \ldots, K_{p_k,q_k}$ for some $k \geq 1$, such that $p_1 q_1 = \cdots = p_k q_k$.

Taking $r = s = 2$ and $t = 4$ in the last theorem we recover (4.14). The original proof in [**33**] is based in successive careful applications of Cauchy Schwarz inequality.

A particularly nice application of (4.14) was observed in Gutman et al. [**17**], where they consider r-regular graphs. A graph G with n vertices is said to be r-regular if the degree of every vertex in the graph is equal to r. In this case G possesses $m = \frac{1}{2}nr$ edges. An upper bound for the energy of a regular graph is readily deduced from McClelland inequality (4.1), i.e.,

$$\mathcal{E}(G) \leq n\sqrt{r}, \tag{4.19}$$

while by applying Koolen-Moulton inequality (4.2), one obtains

$$\mathcal{E}(G) \leq r + \sqrt{(n-1)(nr - r^2)}. \tag{4.20}$$

The following simple lower bound for the energy of regular graphs, in terms of the parameters n and r, follows from (4.14) combined with the fact that the number of closed walks of size 4 in a r-regular graph is bounded from above by nr^3.

THEOREM 4.10 (Gutman et al. [**17**]). *Let G be a regular graph on n vertices, of degree r, $r > 0$. Then*

$$\mathcal{E}(G) \geq n. \tag{4.21}$$

Equality in (4.21) holds if and only if every component of G is isomorphic to the complete bipartite graph $K_{r,r}$.

One could wonder if the very simple inequality $\mathcal{E}(G) \geq n$, holds in general. However, this is not true, a graph G is called *hypoenergetic* if the energy of G is less than the number of vertices of G. In Gutman [**16**] *nonhypoenergetic* graphs (for which $\mathcal{E}(G) \geq n$) are considered.

If we restrict to graphs with no triangles and quadrangles, then the number of closed walks of size 4 is no more than $n(2r^2 - r)$. Thus, in this case we may improve the last bound as follows.

LEMMA 4.11 (Gutman et al. [**17**]). *If G is a regular graph on n vertices, of degree r, $r > 0$ and triangle and quadrangle free, then*

$$\mathcal{E}(G) \geq \frac{nr}{\sqrt{2r-1}}. \tag{4.22}$$

5. Energies of digraphs and matrices

In this section we give the basic definitions of energies of digraphs and matrices. For further details we advice to see [**24, 26, 27**].

5.1. The energy of a digraph. A directed graph (or just digraph) D consists of a non-empty finite set $\mathcal{V}$ of elements called vertices and a finite set $\mathcal{A}$ of ordered pairs of distinct vertices called arcs. Two vertices are called adjacent if they are connected by an arc. Let $A(D)$ be the adjacency matrix of D. It is clear that $A(D)$ is not symmetric in general, thus the eigenvalues can be complex numbers. Recall Coulson's Integral formula for the energy of a graphs

$$\mathcal{E}(G) = \frac{1}{\pi}\int_{-\infty}^{\infty}\left(n - \frac{ix\phi'(G,ix)}{\phi(G,ix)}\right)dx, \tag{5.1}$$

where $\phi(G,x)$ is the characteristic polynomial of graph G, and $\phi'(G,x)$ its derivative.

Motivated by Coulson's integral formula, a generalization of the energy of a graph is given by Peña and Rada in [**26**]. The energy of a digraph D with eigenvalues $z_1, \ldots, z_n$ is given by

$$\mathcal{E}(D) = \sum_{i=1}^{n} |Re(z_i)|, \tag{5.2}$$

where $Re(z_i)$ denotes the real part of z_i. This definition satisfies that for any digraph D we have

$$\mathcal{E}(D) = \frac{1}{\pi}\int_{-\infty}^{\infty}\left(n - \frac{ix\phi'(D,ix)}{\phi(D,ix)}\right)dx. \tag{5.3}$$

It also has the nice property that if $D_1, D_2, \ldots, D_k$ are the strongly connected components of D, then

$$\mathcal{E}(D) = \sum_{i=1}^{k} \mathcal{E}(D_i). \tag{5.4}$$

Thus one can reduce the calculation of the energy of digraphs to strongly connected digraphs, i.e. graphs such that every vertex is reachable from every other vertex.

The analog of McClelland inequality for digraphs was discovered by Rada [**27**].

THEOREM 5.1 ([**27**]). *Let D be a digraph with n vertices, a arcs and let c_2 be the number of closed walks in D of length 2. Then*

$$\mathcal{E}(D) \leq \sqrt{\frac{1}{2}n(a+c_2)}. \tag{5.5}$$

Equality holds in (5.5) if and only if D is the direct sum of $\frac{n}{2}$ copies of K_2, the directed cycle of length 2.

We can see that (5.5) is a generalization of (4.1). In fact, if G is a simple graph with n vertices and m edges we can build its associated symmetric digraph, $\overleftrightarrow{G}$, by declaring that both arcs (x, y) and (y, x) exist in $\overleftrightarrow{G}$ if and only if (x, y) is an edge of G. Then we have that $a = c_2 = 2m$ and $\mathcal{E}(G) = \mathcal{E}(\overleftrightarrow{G}) \leq \sqrt{\frac{1}{2}n(a+c_2)} = \sqrt{2mn}$.

There have been various attempts to give a Koolen-Moulton type inequality for digraphs. On one hand, Gudiño and Rada gave a generalization in [**11**].

THEOREM 5.2 ([**11**]). *Let D be a digraph with n vertices, a arcs and c_2 the number of closed walks in D of length 2. Then*

$$\mathcal{E}(D) \leq \frac{c_2}{n} + \sqrt{(n-1)\left[a - \left(\frac{c_2}{n}\right)^2\right]}. \tag{5.6}$$

Equality holds in (5.6) if and only if D is the empty digraph (i.e., n isolated vertices) or $G = \overleftrightarrow{G}$ where G is one of the following

1. *$G = \frac{n}{2}K_2$,*
2. *$G = K_2$,*
3. *G is a non-complete connected strongly regular graph with two non-trivial eigenvalues both with absolute value $\sqrt{\frac{a-(\frac{c_2}{n})^2}{n-1}}$.*

An application of Theorem 5.2 gives an improvement of Theorem 4.2. More precisely, if G is a graph with n vertices and m edges, the hypothesis $2m \geq n$ can be dropped. In fact, consider $\overleftrightarrow{G}$ the digraph associated to G. Then $\overleftrightarrow{G}$ has n vertices and $a = c_2 = 2m$. Now from Theorem 5.2 we have that

$$\mathcal{E}(G) = \mathcal{E}(\overleftrightarrow{G}) \leq \frac{2m}{n} + \sqrt{(n-1)\left[2m - \left(\frac{2m}{n}\right)^2\right]}.$$

However, inequality (5.6) has the disadvantage that it does not improve (5.5), in general.

Another attempt to generalize Koolen-Moulton' s inequality was made by Ayyaswamy, Balachandran and Gutman [**2**]. They gave the following bound for strongly connected digraphs with $Re(z_1) \geq \frac{a+c_2}{2n} \geq 1$

$$\mathcal{E}(D) \leq \frac{a+c_2}{2n} + \sqrt{(n-1)\left[\frac{a+c_2}{2} - \left(\frac{a+c_2}{2n}\right)^2\right]}. \tag{5.7}$$

Later, Rada [27] gave a stronger inequality, which, however, works only for strongly connected *normal*[3] digraphs

$$\mathcal{E}(D) \leq \frac{a}{n} + \sqrt{(n-1)\left[\frac{a+c_2}{2} - \left(\frac{a}{n}\right)^2\right]}. \tag{5.8}$$

While (5.7), (5.8) do improve (5.5), they have the flaw that they are only valid for digraphs under certain conditions (being normal or satisfying $Re(z_1) \geq \frac{a+c_2}{2n} \geq 1$).

In Section 6 we present a proof of the bound (5.8) for digraphs which are not necessarily normal.

5.2. The energy of Nikiforov. Let $M_{m,n}(\mathbb{C})$ be the set of $m \times n$ matrices with complex entries. For a matrix $A \in M_{m,n}(\mathbb{C})$ we denote by A^* the Hermitian adjoint of A. The singular values $\sigma_1(A) \geq \sigma_2(A) \geq \cdots \geq \sigma_n(A)$ of a matrix A are the square roots of the eigenvalues of AA^*. For any $A \in M_{m,n}(\mathbb{C})$, we call the value $\mathcal{N}(A) = \sigma_1(A) + \ldots + \sigma_n(A)$ the energy of A, which, as noticed by Nikiforov [25], is the nuclear (or trace) norm of A, denoted here by $||A||_*$.

The energy of a matrix generalizes Gutman's definition for the energy of a graph since $\mathcal{E}(G) = \mathcal{N}(A(G))$. However, for a digraph D, the energy of the adjacency matrix of D does not coincide, in general, with the energy of the graph D defined by Peña and Rada [26]. Nikiforov [24] gives an upper bound for $\mathcal{N}(A)$ that generalizes (4.2).

THEOREM 5.3 (Nikiforov [24]). *If $m \leq n$, $A \in M_{m,n}(\mathbb{C})$ is a nonnegative matrix with maximum entry α, and $||A||_1 \geq \alpha n$, then*

$$\mathcal{N}(A) \leq \frac{||A||_1}{\sqrt{mn}} + \sqrt{(n-1)\left[(Tr(AA^*) - \left(\frac{||A||_1}{\sqrt{mn}}\right)^2\right]}, \tag{5.9}$$

where $||A||_1 = \sum_{i=1}^{m}\sum_{j=1}^{n} A_{ij}$.

The proof of (5.9) follows the same lines as the proof of Koolen and Moulton, by bounding from below the largest eigenvalue of A using Rayleigh's principle. As a corollary, by maximizing with respect to $||A||_1$ in (5.9) one gets the following upper bound

$$\mathcal{N}(A) \leq \frac{\alpha(m+\sqrt{m})m}{2}, \tag{5.10}$$

which generalizes (4.1) for the energy of a matrix.

[3] A digraph D is normal if its adjacency matrix, $A = A(D)$, is normal i.e., $A = A^*$.

6. Some relations between energies of graphs, digraphs and matrices

In this section we present some relations between different energies.

6.1. Nikiforov's energy and (n,m)-bipartite graphs. The adjacency matrix A of a bipartite graph whose parts have r and s vertices has the form

$$A = \begin{pmatrix} \mathbf{0}_{r,r} & B \\ B^T & \mathbf{0}_{s,s} \end{pmatrix}, \tag{6.1}$$

where B is a $(0,1)$-matrix of size $r \times s$, and $\mathbf{0}$ represents the zero matrix. Conversely, given a $(0,1)$-matrix B of size $r \times s$, the matrix A as in (6.1) is the adjacency matrix of a bipartite graph. Thus there is a one-to-one correspondence between (r,s)-bipartite graphs and $(0,1)$-matrices of size $r \times s$.

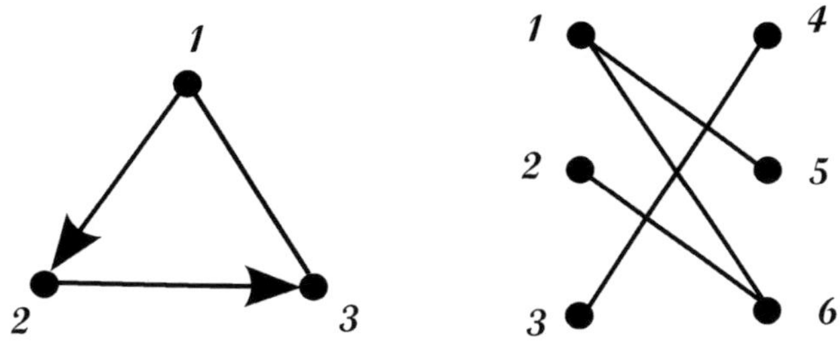

FIGURE 2. A directed graph and its corresponding bipartite graph.

Moreover, it easily follows from the singular value decomposition that for any matrix B, the non-zero eigenvalues of the matrix A in (6.1) are the nonzero singular values of B together with their negatives and thus for any bipartite graph G with adjacency matrix A as in (6.1), one has

$$\mathcal{E}(G) = 2\mathcal{N}(B). \tag{6.2}$$

In other words, studying the energy of (r,s)-bipartite graphs corresponds to studying the Nikiforov's energy of $(0,1)$-matrices of size $r \times s$. The particular case when $r = s = n$ gives a relation between Nikiforov's energy for digraphs and the energy for (n,n)-bipartite graphs.

As an example we can refine inequality (4.10) for bipartite graphs by applying Nikiforov's inequality to B. Indeed, note that for a $(0,1)$-matrix B, we have $||B||_1 = Tr(BB^*) = a$ where a denotes the number of arcs and thus, whenever $a \geq n$, the following inequality holds,

$$\mathcal{N}(B) \leq \frac{a}{\sqrt{rs}} + \sqrt{(r-1)\left[a - \frac{a^2}{rs}\right]}. \tag{6.3}$$

The last inequality together with (6.2) and using the fact that $a(B) = m(G)$, where $A = A(G)$ and B are related by (6.1) yields the following proposition.

PROPOSITION 6.1. *For an (r,s) bipartite graph such that $m \geq min(r,s)$ the following inequality holds.*

$$\mathcal{E}(G) \leq \frac{2m}{\sqrt{rs}} + \sqrt{(min(r,s)-1)\left[4m - \frac{(2m)^2}{rs}\right]}. \tag{6.4}$$

The inequality (6.4) improves the Koolen and Moulton's inequality for bipartite graphs (4.10), since the worst case in the right is $r = s = n/2$, which corresponds to digraphs, and gives exactly (4.10).

Notice that, since $n = r + s$, applying directly inequality (5.9) to A yields,

$$\mathcal{E}(G) \leq \frac{2m}{n} + \sqrt{(n-1)\left[2m - \left(\frac{2m}{n}\right)^2\right]}, \tag{6.5}$$

whenever $m \leq r + s$, which is weaker than (6.4).

On the other hand, we can use the bounds for bipartite graphs to bound from below the energy of a digraph. For instance from the main theorem in [**29**], we can give the analog of inequality (4.14) for digraphs. For this we give an interpretation of the fourth moment of a digraph, i.e., $Tr(AA^*AA^*)$.

An alternating walk of size $2k$ in a digraph is a set of vertices $(v_1, \dots, v_n)$, such that $(v_{2i}, v_{2i+1}) \in E$ and $(v_{2i+2}, v_{2i+1}) \in E$ for all $i = 1, \dots, n$. In other words an alternating walk is a walk that alternates between following the direction of the arrows at odd steps and following the inverse direction of the arrows at even steps. If $v_1 = v_n$ we say that the alternating walk is closed. Thus one arrives to the following theorem.

THEOREM 6.2. *Let D be a digraph of order n with $a \geq 1$ arcs. Then*

$$\mathcal{N}(A(D)) \geq a\sqrt{\frac{a}{q}},$$

where q denotes the number of alternating walks of size 4 in the digraph D.

PROOF. Let $B = A(D)$ be the adjacency matrix of the digraph D and denote by H the (n,n)-bipartite graph with matrix A, as in (6.1). Theorem 1.2 in [**29**] says that for a bipartite graph G we have that

$$\mathcal{E}(G) \geq 2m\sqrt{\frac{m}{m_4}},$$

where $m_4 = M_4^*(G)$. The result follows by (6.2), since $q = m_4$ and $m(H) = a(D)$. □

We would like to mention that the idea of looking at A when trying to say something about B is often used in Random Matrix Theory. This is known as the hermitian reduction method and A is called the hermitianization of B. Also, in [**24**] the author notices that Wigner's semicircle law implies that if we take a uniformly random graph with n vertices, then the energy of this random graph approaches

$$n^{3/2}\Big(\int_1^1 |x|\sqrt{1-x^2}dx\Big) = \frac{4}{3\pi}n^{3/2}. \tag{6.6}$$

It would be interesting if one could obtain the asymptotic behavior of the Nikiforov's energy of a random digraph on n vertices or, more generally, of a random $(0,1)$-matrix of size $m \times n$. Computer simulations suggest that this quantity is of order $n^{3/2}$.

6.2. Bounds for the energy of a digraph. There is a priori not an easy way to relate the energy of a digraph to the energy of a selfadjoint matrix as we did in the last section for the energy of Nikiforov. However in this section we give an upper bound for the energy of a graph in terms of Nikiforov's energy of a matrix. The results presented here are new.

The first step is to compare the quantities $\sum_i |Re\lambda_i(A)|$ and $\sum_i \lambda_i(|ReA|)$, where for a matrix M, $ReM = (M + M^*)/2$ denotes the real part of M and $\lambda_i(M)$ denotes the i'th eigenvalue of M, arranged in decreasing order with respect to the real part.

In the case when A is normal ($A^*A = AA^*$), by functional calculus we have the equality $\sum_i |Re\lambda_i(A)| = \sum_i \lambda_i(|ReA|)$. Thus calculating the energy of a digraph D with normal adjacency matrix A amounts to calculate the nuclear norm of ReA since in this case

$$\mathcal{E}(D) = \mathcal{N}(ReA) = ||ReA||_*. \tag{6.7}$$

In general, $\sum_i |Re\lambda_i(A)|$ need not be equal to $\sum_i \lambda_i(|ReA|)$. However, one may use the following theorem by Ky Fan, to show that $\sum_i |Re\lambda_i(A)| \leq \sum_i \lambda_i(|ReA|)$, (see Theorem III.5.3 in [**3**]).

PROPOSITION 6.3. *(Ky Fan) Let A be a matrix with eigenvalues $\lambda_1, \dots, \lambda_n$ such that $Re(\lambda_1) \geq \cdots \geq Re(\lambda_n)$ and let $\mu_1 \geq \cdots \geq \mu_n$ be the eigenvalues of ReA. For every $k \in \{1, \dots, n\}$, we have*

$$\sum_{i=1}^{k} Re(\lambda_j) \leq \sum_{i=1}^{k} \mu_j. \tag{6.8}$$

The following theorem gives an upper bound for the energy of a digraph in terms of the Nikiforov's energy.

THEOREM 6.4. *For any digraph D with adjacency matrix A we have*

$$\mathcal{E}(D) \leq \mathcal{N}(ReA) \leq \mathcal{N}(A). \tag{6.9}$$

PROOF. Let k be such that $Re(\lambda_1) \geq \cdots \geq Re(\lambda_k) > 0$ and $0 \geq Re(\lambda_{k+1}) \geq \cdots \geq Re(\lambda_n)$. Then $\mathcal{E}(D) = \sum_{i=1}^{k} Re(\lambda_i) - \sum_{i=k+1}^{n} Re(\lambda_i)$. Since the digraph has no loops then its trace is 0 and then $\sum_{i=1}^{k} Re(\lambda_i) = 0$ or $\sum_{i=1}^{k} Re(\lambda_i) = -\sum_{i=k+1}^{n} Re(\lambda_i)$, which implies that $\mathcal{E}(D) = 2\sum_{i=1}^{k} Re(\lambda_i)$. In a similar manner if $\mu_1, \dots, \mu_j$ are the positive eigenvalues of ReA, then $\mathcal{N}(ReA) = 2\sum_{i=1}^{j} \mu_i$ and hence

$$\mathcal{N}(ReA) = 2\sum_{i=1}^{j} \mu_i \geq 2\sum_{i=1}^{k} \mu_i \geq 2\sum_{i=1}^{k} Re(\lambda_k) = \mathcal{E}(D), \tag{6.10}$$

where we used in the last inequality Ky Fan's inequality. This proves the first inequality. The second inequality is well known and follows from Cauchy-Schwarz inequality for matrices. □

REMARK 6.5. We learned about the inequality $\mathcal{E}(D) \leq \mathcal{N}(A)$ from Professor Rada while he was visiting CIMAT.

As a direct corollary we get the McClelland inequality for digraphs proved in Rada [**27**].

Corollary 6.6. *For any digraph D we have the upper bound*

$$\mathcal{E}(D) \leq \sqrt{\frac{n(a+c_2)}{2}}, \tag{6.11}$$

where a denotes the number of arcs of D and c_2 is the number of closed walks in D of length 2.

Proof. A direct application of Cauchy-Schwarz inequality gives

$$\mathcal{N}(|ReA|)^2 \leq nTr[(ReA)^2] = n\Big(\frac{a+c_2}{2}\Big), \tag{6.12}$$

from where we get $\mathcal{E}(D) \leq \mathcal{N}(|ReA|) \leq \sqrt{\frac{n(a+c_2)}{2}}$. □

Now we are in position to give a new Koolen-Moulton type inequality for the energy of a digraph.[4]

Theorem 6.7. *For any digraph D such that $a \geq n$, we have*

$$\mathcal{E}(D) \leq \frac{a}{n} + \sqrt{(n-1)\left[\frac{a+c_2}{2} - \left(\frac{a}{n}\right)^2\right]}. \tag{6.13}$$

Proof. We bound the quantity $\mathcal{N}(ReA)$ using Nikiforov's inequality for the energy of a matrix.

$$\mathcal{N}(ReA) \leq \frac{||ReA||_1}{n} + \sqrt{(n-1)\left[Tr((ReA)^2) - \left(\frac{||ReA||_1}{n}\right)^2\right]}. \tag{6.14}$$

Since $||ReA||_1 = a$ and $Tr((ReA)^2) = (a+c_2)/2$ using Theorem 6.4 we get the desired inequality. □

Remark 6.8.
(1) While the inequality $a \geq n$ may seem restrictive, if $a < n$, then G is not strongly connected, and thus one can apply (5.4) to bound the energy of each of the strongly connected components.
(2) To see that (6.13) improves the inequalities (5.5), (5.7) and (5.6) one proceeds as follows. The conditions for these inequalities to hold ($c_2 \geq \sqrt{an}$ or being strongly connected) imply that $a \geq n$, and thus they satisfy the hypothesis of Theorem 6.7. On the other hand, the function $F(x,y) = x + \sqrt{(n-1)(y-x^2)}$ is decreasing in x for $\sqrt{\frac{y}{n}} \leq x \leq \sqrt{y}$ and increasing in y for $y \geq 0$. Thus, since $a \geq c_2$
$$F(\frac{c_2}{n}, a) \geq F(\frac{a+c_2}{2n}, \frac{a+c_2}{2}) \geq F(\frac{a}{n}, \frac{a+c_2}{2}).$$
This gives the desired inequalities.

Finally, if we specialize to graphs with no cycles of size 2, then we get a much better bound since in this case $\alpha = 2$ in (5.9) and thus we get

$$\mathcal{E}(D) \leq \frac{a}{n} + \sqrt{(n-1)\left[\frac{a}{4} - \left(\frac{a}{n}\right)^2\right]}. \tag{6.15}$$

[4] While this paper was being written, Professor Rada communicated us of another proof of this inequality.

While there are known lower bounds for the energy of a digraph it is not clear how to apply the ideas in last section to bound from below the energy of a digraph unless D is normal.

7. Energy of a vertex

In this section we introduce the definition of the energy of a vertex and derive some properties related to the energy of graph. To the best of our knowledge this definition is new. Moreover, the approach given here to treat the energy of a graph has not been used before. We believe that this could lead to a better understanding of the energy of a graph.

One nice facet of Non Commutative Probability is the fact that the linear functional does not need to be unique and, moreover, there is an interplay between different functionals. Remember that for every $i \in \{1, \ldots, n\}$ the pair (M_n, ϕ_i) is a $*$-non-commutative probability space where ϕ_i is positive and tracial. In this section we want to exploit this fact by decomposing the usual trace as the sum of the linear functionals ϕ_i,

$$Trace(|A(G)|) = \phi_1(|A(G)|) + \cdots + \phi_n(|A(G)|), \tag{7.1}$$

and then analyze each of the linear functionals by its own. For doing this, we define the energy of a vertex in G as follows.

DEFINITION 7.1. Let $G = (V, E)$ be a graph with vertices $v_i, ..., v_n$ and with adjacency matrix $A \in M_n(\mathbb{C})$. We define the energy of the vertex v_i with respect to G by the formula

$$\mathcal{E}_G(v_i) = \phi_i(|A|), \quad \text{for } i = 1, \ldots, n, \tag{7.2}$$

where $|A| = (AA^*)^{1/2}$.

It follows that we can calculate the energy of a graph (or more generally the energy of Nikiforov) by adding the individual energies of the vertices of G,

$$\mathcal{E}(G) = \mathcal{E}_G(v_1) + \cdots + \mathcal{E}_G(v_n). \tag{7.3}$$

The energy of a vertex should be understood as the contribution of this vertex to the energy of the graph, in terms on how it interacts with other vertices. It is easily seen that the energy of a vertex v only depends on the vertices that are in the same component as v. To be more precise, $\mathcal{E}_G(v_i) = \mathcal{E}_{C(v_i)}(v_i)$, where $C(v_i)$ is the connected component of v_i. In particular, one recovers the (trivial) fact that if v is an isolated vertex then it does not contributes to the energy.

Since ϕ_i is positive we may use the inequalities (3.2) and (3.3) from Section 3 which read as follows: For all $i \in \{1, \ldots, n\}$

$$\phi_i(|A|)^2 < \phi_i(|A|^2). \tag{7.4}$$

and

$$\phi_i(|A|) > \phi_i(A^2)^{3/2}/\phi_i(A^4)^{1/2}. \tag{7.5}$$

Thus we arrive to the following improvement of McClelland inequality and the lower bound given in [**9**] in terms of the fourth moment.

THEOREM 7.2. *For a graph G with vertices $v_1, ..., v_n$ with degrees $d_1, ..., d_n$ we have*

$$\frac{(M_2^*)^2}{\sqrt{M_2^* M_4^*}} \leq \sum_i \frac{\phi_i(A^2)^{3/2}}{\phi_i(A^4)^{1/2}} \leq \mathcal{E}(G) \leq \sum_{i=1}^n \sqrt{d_i} \leq \sqrt{2mn}. \tag{7.6}$$

The equality holds in (7.6) if and only if G is a regular graph.

PROOF. The lower bound

$$\sum_i \frac{\phi_i(A^2)^{3/2}}{\phi_i(A^4)^{1/2}} \leq \mathcal{E}(G),$$

follows by adding up (7.5) for all i. Similarly the upper bound

$$\mathcal{E}(G) \leq \sum_{i=1}^n \sqrt{d_i}, \tag{7.7}$$

follows from and the fact that the $\phi_i(|A|)^2 \leq \phi_i(|A|^2) = \phi_i(A^2) = d_i$.

The last inequality $\sum_{i=1}^n \sqrt{d_i} \leq \sqrt{2mn}$ is a direct application of Cauchy-Schwarz (4.2) for the vectors $(d_1, ..., d_n)$ and $(1, ..., 1)$ since $\sum_{i=1}^n d_i = 2m$.

Finally the first inequality is proved as follows. Let $a_i = \phi_i(A^2)$ and $b_i = \phi_i(A^4)$ then applying usual Hölder inequality for $p = 3/2$ and $q = 3$ to the vectors $X = (a_1/b_1^{1/3}, \ldots, a_n/b_n^{1/3})$ and $V = (b_1^{1/3}, \ldots, b_n^{1/3})$ we obtain

$$\sum_{i=1}^n a_i = \sum_{i=1}^n \left(\frac{a_i}{b_i^{1/3}}\right) b_i^{1/3} \leq \left(\sum_{i=1}^n \frac{a_1^{3/2}}{b_i^{1/2}}\right)^{2/3} \left(\sum_{i=1}^n b_i\right)^{1/3}. \tag{7.8}$$

Since $\sum_i a_i = Tr(A^2) = M_2^*$, and $\sum_i b_i = Tr(A^4) = M_4^*$ by rearranging we get the desired result. □

References

[1] S. Akbari, E. Ghorbani, and S. Zare, *Some relations between rank, chromatic number and energy of graphs*, Discrete Math. **309** (2009), no. 3, 601–605. MR2499013

[2] S. K. Ayyaswamy, S. Balachandran, and I. Gutman, *Upper bound for the energy of strongly connected digraphs*, Appl. Anal. Discrete Math. **5** (2011), no. 1, 37–45. MR2809031

[3] R. Bhatia, *Matrix analysis*, Graduate Texts in Mathematics, vol. 169, Springer-Verlag, New York, 1997. MR1477662

[4] A. E. Brouwer and W. H. Haemers, *Spectra of graphs*, Universitext, Springer, New York, 2012. MR2882891

[5] Caporossi, G., Cvetković, D., Gutman, I., & Hansen, B., (1999). Variable neighborhood search for extremal graphs. 2. Finding graphs with extremal energy. Journal of Chemical Information and Computer Sciences, 39(6), 984-996.

[6] L. Collatz and U. Sinogowitz, *Spektren endlicher Grafen* (German), Abh. Math. Sem. Univ. Hamburg **21** (1957), 63–77. MR0087952

[7] Coulson, C., (1940). On the calculation of the energy in unsaturated hydrocarbon molecules. Mathematical Proceedings of the Cambridge Philosophical Society, 36(2), 201-203.

[8] R. Cruz, H. Giraldo, and J. Rada, *An upper bound for the energy of radial digraphs*, Linear Algebra Appl. **442** (2014), 75–81. MR3134352

[9] J. A. de la Peña and L. Mendoza, *Moments and π-electron energy of hexagonal systems in 3-space*, MATCH Commun. Math. Comput. Chem. **56** (2006), no. 1, 113–129. MR2312474

[10] J. A. de la Peña, L. Mendoza, and J. Rada, *Comparing momenta and π-electron energy of benzenoid molecules*, Discrete Math. **302** (2005), no. 1-3, 77–84. MR2179637

[11] E. Gudiño and J. Rada, *A lower bound for the spectral radius of a digraph*, Linear Algebra Appl. **433** (2010), no. 1, 233–240. MR2645080

[12] Gutman, I. (1974) Bounds for total π-electron energy. Chemical Physics Letters, 24(2), 283-285.

[13] I. Gutman, *The energy of a graph*, Ber. Math.-Statist. Sekt. Forsch. Graz **100-105** (1978), Ber. No. 103, 22. 10. Steiermärkisches Mathematisches Symposium (Stift Rein, Graz, 1978). MR525890

[14] Gutman, I. (1985). Bounds for total π-electron energy of conjugated hydrocarbons. Zeitschrift für Physikalische Chemie, 266(1), 59-64.

[15] I. Gutman, *The energy of a graph: old and new results*, Algebraic combinatorics and applications (Gößweinstein, 1999), Springer, Berlin, 2001, pp. 196–211. MR1851951

[16] I. Gutman, *On graphs whose energy exceeds the number of vertices*, Linear Algebra Appl. **429** (2008), no. 11-12, 2670–2677. MR2455521

[17] I. Gutman, S. Zare Firoozabadi, J. A. de la Peña, and J. Rada, *On the energy of regular graphs*, MATCH Commun. Math. Comput. Chem. **57** (2007), no. 2, 435–442. MR2310776

[18] I. Gutman, E. Gudiño, and D. Quiroz, *Upper bound for the energy of graphs with fixed second and fourth spectral moments*, Kragujevac J. Math. **32** (2009), 27–35. MR2584237

[19] A. Hora and N. Obata, *Quantum probability and spectral analysis of graphs*, Theoretical and Mathematical Physics, Springer, Berlin, 2007. With a foreword by Luigi Accardi. MR2316893

[20] J. H. Koolen and V. Moulton, *Maximal energy graphs*, Adv. in Appl. Math. **26** (2001), no. 1, 47–52. MR1806691

[21] J. H. Koolen and V. Moulton, *Maximal energy bipartite graphs*, Graphs Combin. **19** (2003), no. 1, 131–135. MR1974375

[22] X. Li, Y. Shi, and I. Gutman, *Graph energy*, Springer, New York, 2012. MR2953171

[23] McClelland, B. J. (1971). Properties of the Latent Roots of a Matrix: The Estimation of p-Electron Energies. The Journal of Chemical Physics, 54(2), 640-643.

[24] V. Nikiforov, *The energy of graphs and matrices*, J. Math. Anal. Appl. **326** (2007), no. 2, 1472–1475. MR2280998

[25] V. Nikiforov, *On the sum of k largest singular values of graphs and matrices*, Linear Algebra Appl. **435** (2011), no. 10, 2394–2401. MR2811124

[26] I. Peña and J. Rada, *Energy of digraphs*, Linear Multilinear Algebra **56** (2008), no. 5, 565–579. MR2437906

[27] J. Rada, *The McClelland inequality for the energy of digraphs*, Linear Algebra Appl. **430** (2009), no. 2-3, 800–804. MR2473186

[28] J. Rada, *Bounds for the energy of normal digrahs*, Linear Multilinear Algebra **60** (2012), no. 3, 323–332. MR2886727

[29] J. Rada and A. Tineo, *Upper and lower bounds for the energy of bipartite graphs*, J. Math. Anal. Appl. **289** (2004), no. 2, 446–455. MR2026917

[30] D. Voiculescu, *Limit laws for random matrices and free products*, Invent. Math. **104** (1991), no. 1, 201–220. MR1094052

[31] D. V. Voiculescu, K. J. Dykema, and A. Nica, *Free random variables*, CRM Monograph Series, vol. 1, American Mathematical Society, Providence, RI, 1992. A noncommutative probability approach to free products with applications to random matrices, operator algebras and harmonic analysis on free groups. MR1217253

[32] B. Zhou, *Energy of a graph*, MATCH Commun. Math. Comput. Chem. **51** (2004), 111–118. MR2063930

[33] B. Zhou, I. Gutman, J. A. de la Peña, J. Rada, and L. Mendoza, *On spectral moments and energy of graphs*, MATCH Commun. Math. Comput. Chem. **57** (2007), no. 1, 183–191. MR2293903

Centro de Investigación en Matemáticas, Guanajuato, Mexico
Email address: octavius@cimat.mx

Centro de Investigación en Matemáticas, Guanajuato, Mexico
Email address: ojuarez@cimat.mx

Contemporary Mathematics
Volume **709**, 2018
http://dx.doi.org/10.1090/conm/709/14289

The Atiyah-Singer cobordism invariance and the tangent groupoid

Paulo Carrillo Rouse

Abstract. We give a new proof of the cobordism invariance of Atiyah-Singer's analytic index morphism in K-theory by means of Connes' tangent groupoid. The new proof allows to refine Carvalho's previous K-theory proof in the sense that we prove the invariance for symbols which are equivalent via noncommutative symbols coming from fully elliptic b-pseudodifferential operators giving as a byproduct a generalization of Moroianu's analytic interpretation of the cobordism invariance.

1. Introduction

The first proof of the famous Atiyah-Singer index theorem used the cobordism invariance of the index of the twisted signature operator on a closed smooth manifold, [**AS63**]. In [**Car05**], Carvalho generalized the cobordism index invariance for all pseudodifferential elliptic operators using topological K-theory methods. The main point of her paper was to obtain the cobordism invariance as a consequence of the compatibility of index morphisms with Gysin maps in K-theory. The methods developed in [**Car05**] apply as well for non compact manifolds for which Carvalho interpreted the analytic index and the cobordism invariance in terms of operators which act by multiplication at infinity, in particular she obtained, as a consequence, the extension to this setting of the K-theoretical Atiyah-Singer index theorem.

The purpose of this note is to give a new proof of Carvalho's result using deformation groupoids techniques. In fact the new proof will allow us to refine the symbol cobordism invariance defined in [**Car05**] definition 2.1 by allowing cobordisms coming from the noncommutative tangent space associated to a manifold with boundary.

The use of groupoids in Index theory can be traced back to Connes and Connes-Skandalis, see Connes' book [**Con94**] chapter II for a good survey. It allows in one side to deal in a unified way with different kind of geometric situations, families, foliations, coverings, actions, singular manifolds and many others, and in the other hand it can be used to highly simplify the proofs as for example Connes' proof of Atiyah-Singer index theorem using the tangent groupoid.

2010 *Mathematics Subject Classification.* Primary 58H05, 19K56; Secondary 58B34.

Key words and phrases. Index theory, Lie groupoids, K-theory, noncommutative differential geometry.

Supported by ANR-14-CE25-0012-01 and Max Planck Institut für Mathematik at Bonn. The author is very grateful to these institutions for their support.

Besides giving a new proof and a refinement of a known theorem, the objective of this paper is to survey the groupoids techniques that can certainly, and will, be applied in higher or more singular situations.

The paper is organized as follows. In section 2 we revisit the deformation to the normal cone functor for which the Connes tangent groupoid is a particular example. Deformations of this kind appear also later as orbit spaces from which the indices can be computed. In section 3 we recall how the analytic index morphism may be constructed by means of the tangent groupoid of a manifold, we also recall in these terms Connes' construction of the topological index morphism and the Atiyah-Singer index theorem. In section 4 we prove our cobordism invariance result, for this we define first the cobordism relation via the K-theory of a Lie groupoid called the noncommutative tangent space of a manifold with boundary. In section 5 we discuss the analytic interpretation of the cobordism invariance, in particular we recall that in the compact case the K-theory of the noncommutative tangent space encodes noncommutative symbols of b-pseudodifferential operators, giving thus a b-interpretation of Moroianu's cobordism invariance in terms of cusps operators in [**Mor06**].

Acknowledgments: The author wishes to thank the referee for useful remarks and corrections that helped to improve the paper presentation.

2. Connes tangent groupoid and the deformation to the normal cone functor

Connes tangent groupoid is a particular case of a geometric construction that we describe here.

Let M be a C^∞ manifold and $X \subset M$ be a C^∞ submanifold. We denote by $N(M,X)$ the normal bundle to X in M. We define the following set

$$D(M,X) := N(M,X) \bigsqcup M \times \mathbb{R}^* \tag{2.1}$$

We recall how to define a C^∞-structure in $D(M,X)$.

Let us first consider the case where $M = \mathbb{R}^p \times \mathbb{R}^q$ and $X = \mathbb{R}^p \times \{0\}$ (here we identify X canonically with $\mathbb{R}^p$). We denote by $q = n - p$ and by $D(n,p)$ for $D(\mathbb{R}^n, \mathbb{R}^p)$ as above. In this case we have that $D(n,p) = \mathbb{R}^p \times \mathbb{R}^q \times \mathbb{R}$ (as a set). Consider the bijection $\psi : \mathbb{R}^p \times \mathbb{R}^q \times \mathbb{R} \to D(n,p)$ given by

$$\psi(x,\xi,t) = \begin{cases} (x,\xi,0) & \text{if } t = 0 \\ (x,t\xi,t) & \text{if } t \neq 0. \end{cases} \tag{2.2}$$

We can consider the C^∞-structure on $D(n,p)$ induced by this bijection.

Now, in the general case, a local chart $(\mathcal{U},\phi)$ in M is said to be a X-slice if

1) $\phi : \mathcal{U} \to U \subset \mathbb{R}^p \times \mathbb{R}^q$ is a diffeomorphsim.
2) If $V = U \cap (\mathbb{R}^p \times \{0\})$, then $\phi^{-1}(V) = \mathcal{U} \cap X$, denoted by $\mathcal{V}$.

With this notation, $D(U,V) \subset D(n,p)$ as an open subset. We may define a function

$$\tilde{\phi} : D(\mathcal{U},\mathcal{V}) \to D(U,V) \tag{2.3}$$

in the following way: For $x \in \mathcal{V}$ we have $\phi(x) \in \mathbb{R}^p \times \{0\}$. If we write $\phi(x) = (\phi_1(x), 0)$, then

$$\phi_1 : \mathcal{V} \to V \subset \mathbb{R}^p$$

is a diffeomorphism. We set $\tilde{\phi}(v,\xi,0) = (\phi_1(v), d_N\phi_v(\xi), 0)$ and $\tilde{\phi}(u,t) = (\phi(u),t)$ for $t \neq 0$. Here $d_N\phi_v : N_v \to \mathbb{R}^q$ is the normal component of the derivative $d\phi_v$ for

$v \in \mathcal{V}$. It is clear that $\tilde{\phi}$ is also a bijection (in particular it induces a C^∞ structure on $D(\mathcal{U}, \mathcal{V})$).

Let us define, with the same notations as above, the following set

$$\Omega_V^U = \{(x, \xi, t) \in \mathbb{R}^p \times \mathbb{R}^q \times \mathbb{R} : (x, t \cdot \xi) \in U\}. \tag{2.4}$$

which is an open subset of $\mathbb{R}^p \times \mathbb{R}^q \times \mathbb{R}$ and thus a C^∞ manifold. It is immediate that $D(U, V)$ is diffeomorphic to Ω_V^U through the restriction of Ψ, used in (2.2). Now we consider an atlas $\{(\mathcal{U}_\alpha, \phi_\alpha)\}_{\alpha \in \Delta}$ of M consisting of $X-$slices. It is clear that

$$D(M, X) = \cup_{\alpha \in \Delta} D(\mathcal{U}_\alpha, \mathcal{V}_\alpha) \tag{2.5}$$

and if we take $D(\mathcal{U}_\alpha, \mathcal{V}_\alpha) \overset{\varphi_\alpha}{\to} \Omega_{V_\alpha}^{U_\alpha}$ defined as the composition

$$D(\mathcal{U}_\alpha, \mathcal{V}_\alpha) \overset{\tilde{\phi_\alpha}}{\to} D(U_\alpha, V_\alpha) \overset{\Psi_\alpha^{-1}}{\to} \Omega_{V_\alpha}^{U_\alpha}$$

then we have (proposition 3.1 in [**CR08**]).

Proposition 2.1. *$\{(D(\mathcal{U}_\alpha, \mathcal{V}_\alpha), \varphi_\alpha)\}_{\alpha \in \Delta}$ is a C^∞ atlas over $D(M, X)$.*

Definition 2.2 (Deformation to the normal cone). Let $X \subset M$ be as above. The set $D(M, X)$ equipped with the C^∞ structure induced by the atlas of X-slices is called the deformation to the normal cone associated to the embedding $X \subset M$.

One important feature about the deformation to the normal cone is the functoriality. More explicitly, let $f : (M, X) \to (M', X')$ be a C^∞ map $f : M \to M'$ with $f(X) \subset X'$. Define $D(f) : D(M, X) \to D(M', X')$ by the following formulas:

1) $D(f)(m, t) = (f(m), t)$ for $t \neq 0$,
2) $D(f)(x, \xi, 0) = (f(x), d_N f_x(\xi), 0)$, where $d_N f_x$ is by definition the map

$$(N(M, X))_x \overset{d_N f_x}{\longrightarrow} (N(M', X')_{f(x)}$$

induced by $T_x M \overset{df_x}{\longrightarrow} T_{f(x)} M'$.

Then we have, (proposition 3.6 in [**CR08**]),

Proposition 2.3. *$D(f) : D(M, X) \to D(M', X')$ is a C^∞-map. In other terms, the deformation to the normal cone construction defines a functor*

$$D : \mathcal{C}_2^\infty \longrightarrow \mathcal{C}^\infty, \tag{2.6}$$

where $\mathcal{C}^\infty$ is the category of C^∞-manifolds and $\mathcal{C}_2^\infty$ is the category of pairs of C^∞-manifolds.

We will apply last construction to Lie groupoids, we briefly recall the definition of such an object:

Definition 2.4. A Lie *groupoid* consists of the following data: two C^∞-manifolds G and M, and differentiable maps

(1) $s, r : G \to M$ called the source map and target map respectively and assumed to be submersions,
(2) $m : G^{(2)} \to G$ called the product map (where $G^{(2)} = \{(\gamma, \eta) \in G \times G : s(\gamma) = r(\eta)\}$),

together with two additional maps, $u : M \to G$ (the unit map) and $i : G \to G$ (the inverse map), such that, if we denote $m(\gamma, \eta) = \gamma \cdot \eta$, $u(x) = x$ and $i(\gamma) = \gamma^{-1}$, we have

(i). $r(\gamma \cdot \eta) = r(\gamma)$ and $s(\gamma \cdot \eta) = s(\eta)$.
(ii). $\gamma \cdot (\eta \cdot \delta) = (\gamma \cdot \eta) \cdot \delta$, $\forall \gamma, \eta, \delta \in G$ whenever this makes sense.
(iii). $\gamma \cdot x = \gamma$ and $x \cdot \eta = \eta$, $\forall \gamma, \eta \in G$ with $s(\gamma) = x$ and $r(\eta) = x$.
(iv). $\gamma \cdot \gamma^{-1} = u(r(\gamma))$ and $\gamma^{-1} \cdot \gamma = u(s(\gamma))$, $\forall \gamma \in G$.

For simplicity, we denote a groupoid by $G \rightrightarrows M$. A strict morphism f from a groupoid $H \rightrightarrows M'$ to a groupoid $G \rightrightarrows M$ is given by maps in

$$\begin{array}{ccc} H & \xrightarrow{f} & G \\ \downdownarrows & & \downdownarrows \\ M' & \xrightarrow{f} & M \end{array}$$

which preserve the groupoid structure, i.e., f commutes with the source, target, unit, inverse maps, and respects the groupoid product in the sense that $f(h_1 \cdot h_2) = f(h_1) \cdot f(h_2)$ for any $(h_1, h_2) \in H^{(2)}$.

Definition 2.5. [Tangent groupoid of a Lie groupoid] Let $G \rightrightarrows M$ be a Lie groupoid. *The tangent groupoid* associated to G is the Lie groupoid[1] obtained by applying the Deformation to the normal cone functor to the groupoid G (restricted to the closed interval $[0,1]$), that is

$$D(G, M) \rightrightarrows D(M, M) = M \times [0,1]$$

with structural maps given by $D(s), D(t)$ for source and target, $D(m)$ for the product, and $D(u)$, $D(i)$ for the unit and inverse maps. Usually it is denoted by

$$G^{tan} \rightrightarrows M \times [0,1].$$

The vector bundle $AG = N(G, M) \to M$ is called the Lie algebroid of G (In this paper we will not use its Lie algebroid structure so we will not enter into these details). It can be identified with the vector bundle

$$AG = Ker\, ds|_M = \bigsqcup_{x \in M} T_x G_x. \tag{2.7}$$

3. Index theory and deformation groupoids

For a Lie groupoid $G \rightrightarrows M$ there is a naturally associated C^*-algebra, denoted by $C^*(G)$, given as a maximal completion[2] of the convolution algebra of smooth compactly supported functions $C_c^\infty(G)$ (complex valued functions). For more details on these algebras and their properties the reader might consult the the very nice textbook [**Kha13**] by Khalkali.

Notation: In the sequel we will be using the following notation for the K-theory groups of the C^*-algebra of a groupoid.

$$K^*(G) := K_*(C^*(G)), \tag{3.1}$$

and in particular

$$K^*_{top}(X) = K_*(C_0(X)) \tag{3.2}$$

for topological K-theory of a space when the groupoid is a space (groupoid of units).

[1] It is a Lie groupoid because of the functoriality of the deformation to the normal cone.

[2] All the groupoids treated in this paper are amenable and hence we are not require to enter into a discussion about different kinds of completions.

Let X be a smooth manifold without boundary. Consider the pair groupoid

$$X \times X \rightrightarrows X$$

and its tangent groupoid

$$G_X := (X \times X)^{tan} \rightrightarrows X \times [0,1].$$

It comes equipped with a short exact sequence of C^*-algebras

$$0 \longrightarrow C^*(X \times X \times (0,1]) \longrightarrow C^*(G_X) \xrightarrow{e_0} C^*(TX) \longrightarrow 0 \tag{3.3}$$

where e_0 stands for the evaluation at zero morphism. Since the C^*-algebra $C^*(X \times X \times (0,1])$ is obviously contractible and the K-theory of C^*-algebras functor is homotopy invariant we get that the morphism induced by e_0 (which will be denoted the same) in K-theory is an isomorphism. We can now consider the $K-$theory morphism

$$Ind_X := e_1 \circ (e_0)^{-1} : K^0(TX) \to \mathbb{Z} \tag{3.4}$$

where $e_1 : K^0(G_X) \to K^0(X \times X) \cong \mathbb{Z}$ is induced from the evaluation at $t = 1$, $e_1 : C^*(G_X) \to C^*(X \times X) \cong \mathcal{K}(L^2(X))$ (the algebra of compact operators). The morphism Ind_X is the analytic index morphism of X but we will leave this justification for the final section, for the moment we will refer to it as the analytic index of X even if the manifold is not compact (again we will discuss at the end of the paper what the choice of our terminology means). We will briefly sketch how the above morphism can be computed topologically.

The topological index and the Atiyah-Singer index theorem: Consider an embedding $X \overset{i}{\hookrightarrow} \mathbb{R}^N$ with N even. It induces a groupoid morphism

$$G_X \xrightarrow{h} \mathbb{R}^N$$

defined by

$$h(x,y,t) := \frac{i(x) - i(y)}{t}$$

for $t \neq 0$ and by

$$h(x,V) := d_x i(V)$$

for $t = 0$. The induced semidirect product groupoid $G_X \ltimes_h \mathbb{R}^N$ is easily seen to be free and proper and hence Morita equivalent to its orbit space. A direct computation allows to show that the orbit space is diffeomorphic to the deformation to the normal cone associated to the embedding, *i.e.*,

$$Orb(G_X \ltimes_h \mathbb{R}^N) \cong D(\mathbb{R}^N, X) = N(\mathbb{R}^N, X) \bigsqcup \mathbb{R}^N \times (0,1].$$

One can consider the morphism induced by this orbit space, that is the $K-$theory morphism given by the following compositions
(3.5)

$$Ind_X^{top} : K^0_{top}(T^*X) \xrightarrow[\cong]{T} K^0_{top}(N(\mathbb{R}^N, X)) \xleftarrow[e_0]{\cong} K^0_{top}(D(\mathbb{R}^N, X)) \xrightarrow{e_1} K^0_{top}(\mathbb{R}^N) \xrightarrow{B^{-1}} \mathbb{Z}.$$

where T stands for the Thom isomorphism, B for the Bott isomorphism, and the e_i are induced from the respective inclusions at $t = 0$ and $t = 1$. The morphism above is exactly the Atiyah-Singer topological index morphism as it is immediate from the structure of $D(\mathbb{R}^N, X)$ that the morphism $e_1 \circ (e_0)^- - 1 = j! : K^0_{top}(N(\mathbb{R}^N, X)) \to K^0_{top}(\mathbb{R}^N)$ where $j!$ is just given by extension by zero from the open subset $N(\mathbb{R}^N, X)$ to $\mathbb{R}^N$.

In [**Con94**] chapter II, Connes shows using the tangent groupoid the Atiyah-SInger index theorem

$$Ind_X = Ind_X^{top}. \tag{3.6}$$

Connes' proof that the two morphisms above are equal, or see also theorem 4.4 in [**CRLM14**], works even if X is not closed, the analytic interpretation is however different as we will discuss in the last section.

4. Cobordism invariance of the index

For defining the cobordism relation we need to introduce certain groupoids associated to manifolds with boundary.

Let W a manifold with boundary $\partial W = X$. We will take from now a defining function of the boundary $\rho : W \to \mathbb{R}^+$, and consider the b-groupoid

$$\Gamma_b(W) := \{(x, y, \lambda) \in W \times W \times \mathbb{R} : \rho(x) = e^\lambda \cdot \rho(y)\} \tag{4.1}$$

as a subgroupoid of the product groupoid $W \times W \times \mathbb{R} \rightrightarrows W$, $\mathbb{R}$ considered as an additive group (and hence groupoid). This groupoid inherits a continuous families groupoid structure. It was initially introduced in [**Mon03**] by Monthubert to give a groupoid description of the $b-$pseudodifferential calculus.

We might consider the tangent groupoid

$$\Gamma_b(W)^{tan} \rightrightarrows W \times [0, 1] \tag{4.2}$$

and a particular restriction

$$\Gamma_b(W)^{Fred} := \Gamma_b(W)^{tan}|_{W_F} \rightrightarrows W_F \tag{4.3}$$

where $W_F := W \times [0, 1] \setminus (X \times \{1\})$. We call the last open subgroupoid the *Fredholm groupoid of* W for reasons explained below. Let $\overset{\circ}{W}$ the interior of W, by definition $\Gamma_b(W)^{Fred}$ contains $\overset{\circ}{W} \times \overset{\circ}{W} \times (0, 1] \rightrightarrows \overset{\circ}{W} \times (0, 1]$ as a saturated open subgroupoid with closed complement the closed saturated subgroupoid

$$T_{nc}W \rightrightarrows W_\partial,$$

where we are denoting $T_{nc}W := \Gamma_b(W)^{Fred} \setminus \overset{\circ}{W} \times \overset{\circ}{W} \times (0, 1]$ and $W_\partial := W_F \setminus \overset{\circ}{W} \times (0, 1]$. The groupoid $T_{nc}W$ is called *the noncommutative tangent space of* W, we will mention something about this in the last section.

By construction of the noncommutative tangent space $T_{nc}W$, if one takes the groupoid restriction to $X \times \{0\} \subset W_\partial$ one gets the closed subgroupoid $TX \times \mathbb{R} \rightrightarrows X$, and so at the level of C^*-algebras one has an induced restriction morphism

$$C^*(T_{nc}W) \xrightarrow{r} C^*(TX \times \mathbb{R}) \cong C_0(T^*X \times \mathbb{R}) \tag{4.4}$$

that induces a morphism in K-theory

$$K^1(T_{nc}W) := K^1(C^*(T_{nc}W)) \xrightarrow{r} K^1(C_0(T^*X \times \mathbb{R})) \cong K^0_{top}(T^*X). \tag{4.5}$$

DEFINITION 4.1 (b-cobordism). Let X_i, i=0,1, be a couple of manifolds (without boundary), and let $\sigma_i \in K^0_{top}(T^*X_i)$, $i = 0, 1$, be a couple of principal symbol classes. We say that the pairs (X_0, σ_0) and (X_1, σ_1) are b-cobordant if there is a manifold with boundary W and $a \in K^1(T_{nc}W)$ such that

(i) The boundary of W satisfies $\partial W = X_0 \bigsqcup X_1$, and
(ii) $r_0(a) = \sigma_0$ and $r_1(a) = -\sigma_1$ where $r_i : K^1(T_{nc}W) \to K^0_{top}(T^*X_i)$ is the restriction morphism described above.

In this case we write $(X_0, \sigma_0) \sim_b (X_1, \sigma_1)$.

REMARK 4.2. The cobordism relation above refines Carvalho's relation, [**Car05**]. Indeed, we have a commutative diagram

$$\begin{array}{ccc} K^1(T_{nc}W) & \xrightarrow{\qquad\qquad r \qquad\qquad} & K^0_{top}(T^*X) \\ & {\scriptstyle r}\searrow \qquad \nearrow{\scriptstyle r} & \\ & K^1_{top}(T^*W) & \end{array} \tag{4.6}$$

where $K^1(T_{nc}W) \to K^1(T^*W)$ is given by the groupoid restriction to W and $K^1_{top}(T^*W) \to K^0_{top}(T^*X)$ is the topological restriction considered by Carvalho (also obtained by restricting the groupoid TW to X and applying Fourier transforms) and used by her to define a cobordism relation. Now, for reasons explained in the last section (Lescure's result), the morphism $K^1(T_{nc}W) \to K^1(T^*W)$ cannot be surjective (unless $\partial W = \emptyset$) and so our relation is a refinement of Carvalho's one.

We are ready to state and prove our main theorem:

THEOREM 4.3. *Let X be a smooth manifold without boundary. Suppose there is a smooth manifold with boundary W whose boundary is $\partial W = X$, then the morphism*

$$K^1(T_{nc}W) \xrightarrow{r} K^0_{top}(T^*X) \xrightarrow{Ind_X} \mathbb{Z} \tag{4.7}$$

is the zero morphism.

PROOF. Consider $W \xrightarrow{i'_W} \mathbb{R}^N$ an embedding with N an even natural number and $W \xrightarrow{i_W} \mathbb{R}^{N+1}$ the associated neat embedding, $i_W(w) = (i'_W(w), \rho(w))$. In particular we also have an induced embedding $X \xrightarrow{i_X} \mathbb{R}^N$. From the embedding i_W one has a continuous groupoid morphism

$$W \times W \to \mathbb{R}^{N+1}$$

that restricts to a Lie groupoid morphism

$$\Gamma(W) \xrightarrow{h} \mathbb{R}^{N+1}.$$

By functoriality of the tangent groupoid functor we have a morphism, that we will also denote by h,

$$\Gamma(W)^{tan} \xrightarrow{h} \mathbb{R}^{N+1}$$

given by the composition of h^{tan} ($D(h)$ in the notation of the deformation to the normal cone functor) followed by the canonical projection $(\mathbb{R}^{N+1})^{tan} \to \mathbb{R}^{N+1}$. By restriction we get a morphism on the Fredholm groupoid

$$\Gamma(W)^{Fred} \xrightarrow{h} \mathbb{R}^{N+1}$$

that is compatible with its saturated subgroupoids $T_{nc}W$ and $\mathring{W} \times \mathring{W} \times (0,1]$. In [**CRLM14**] proposition 3.6 we showed that the semi-direct product groupoid $\Gamma(W)^{Fred} \ltimes_h \mathbb{R}^{N+1}$ is free and proper, and hence Morita equivalent to its orbit space.

We have an associated K-theory morphism

$$K^1(T_{nc}W) \xrightarrow[CT]{\cong} K^0_{top}(N_{sing}(\mathbb{R}^{N+1}, W)) \xleftarrow[e_0]{\cong} K^0_{top}(B_{sing}(\mathbb{R}^{N+1}, W)) \xrightarrow[e_1]{} K^0_{top}(\mathbb{R}^{N+1}) \tag{4.8}$$

where $B_{sing}(\mathbb{R}^{N+1}, W)$ is the orbit space of $\Gamma(W)^{Fred} \ltimes_h \mathbb{R}^{N+1}$, that after proposition 3.11 in [**CRLM14**] is a C^∞-cobordism between $N_{sing}(\mathbb{R}^{N+1}, W)$, the orbit space of $T_{nc}W \ltimes_h \mathbb{R}^{N+1}$ and $\mathbb{R}^{N+1}$, the orbit space of $\mathring{W} \times \mathring{W} \ltimes_h \mathbb{R}^{N+1}$. The morphism CT stands for the natural Connes-Thom ismorphism $K^1(T_{nc}W) \to K^0(T_{nc}W \ltimes_h \mathbb{R}^{N+1}) \cong K^0_{top}(N_{sing}(\mathbb{R}^{N+1}, W))$, $N+1$ is odd and so CT exchanges the K-groups gradings. The morphism e_0 and e_1 stand for the morphisms induced by the inclusions, the fact that e_0

is an isomorphism is obtained from an easy contractibility argument as in (3.3). Now, by definition

$$\Gamma(W)^{Fred}|_{X\times[0,1]} = G_X^{tan}$$

and

$$h_X := h|_{G_X^{tan}} : G_X^{tan} \to \mathbb{R}^N$$

is the groupoid morphism induced by i_X and whose K-theory morphism

$$K^0(TX) \xrightarrow[\cong]{CT} K^0_{top}(N(\mathbb{R}^N, X)) \xleftarrow[\cong]{e_0} K^0_{top}(D(\mathbb{R}^N, X)) \xrightarrow{e_1} K^0_{top}(\mathbb{R}^N) \tag{4.9}$$

computes the index

$$K^0_{top}(T^*X) \overset{Ind_X}{\longrightarrow} \mathbb{Z}$$

once taking into account the Fourier isomorphism $K^0_{top}(T^*X) \cong K^0(TX)$ and the Bott isomorphism $K^0_{top}(\mathbb{R}^N) \cong \mathbb{Z}$.

The fact that $h_X := h|_{G_X^{tan}}$ transposes at every level to give the following commutative diagram

(4.10)

$$\begin{array}{ccccccc}
K^0(TX) & \xrightarrow[\cong]{CT} & K^0_{top}(N(\mathbb{R}^N, X)) & \xleftarrow[\cong]{e_0} & K^0_{top}(D(\mathbb{R}^N, X)) & \xrightarrow{e_1} & K^0_{top}(\mathbb{R}^N) \\
\uparrow r & & \uparrow r & & \uparrow r & & \uparrow r \\
K^1(T_{nc}W) & \xrightarrow[CT]{\cong} & K^0_{top}(N_{sing}(\mathbb{R}^{N+1}, W)) & \xleftarrow[e_0]{\cong} & K^0_{top}(B_{sing}(\mathbb{R}^{N+1}, W)) & \xrightarrow[e_1]{} & K^0_{top}(\mathbb{R}^{N+1})
\end{array}$$

where the morphisms r stand for the induced morphisms given by restriction. We conclude the proof from the above diagram since N is supposed to be even. □

The following is a classic immediate consequence of a theorem as the one above:

COROLLARY 4.4. *Let X_i, i=0,1, be a couple of manifolds (without boundary), and let $\sigma_i \in K^0(T^*X_i)$, $i = 0, 1$, be a couple of principal symbol classes, if $(X_0, \sigma_0) \sim_b (X_1, \sigma_1)$ are b-cobordant then*

$$Ind_{X_0}(\sigma_0) = Ind_{X_1}(\sigma_1). \tag{4.11}$$

REMARK 4.5. The same tangent groupoid argument applies to give a new proof of the cobordism invariance using Carvalho's cobordism relation, *i.e.* using instead the restriction morphism $K^1_{top}(T^*W) \to K^0(T^*X)$.

5. Analytic interpretation

The reason we have not had insisted too much in the compacity of X is because the results above about the equality of two morphisms holds without this hypothesis. However we have the following analytic interpretations:

For X compact, we know since the seminal work of Atiyah and Singer that $K^0_{top}(T^*X)$ classifies all principal symbol classes of classic pseudodifferential elliptic operators acting between sections of vector bundles and Ind_X computes indeed the Fredholm index (see for example [**MP97**] or Connes Book [**Con94**]). Lescure generalized the above fact for manifolds with boundary, indeed, if W is a compact manifold with boundary it was shown in [**Les09**] Theorem 1 that every element in $K^1(T_{nc}W)$ can be obtained as the class induced by the (noncommutative) principal symbol of a (ungraded) fully elliptic b-pseudodifferential operator of order zero. So our result on the cobordism invariance of the index extends Moroianu's analytic interpretation, [**Mor06**], for Carvalho's theorem in this setting.

For X non compact, Carvalho showed in her paper that $K^0_{top}(T^*X)$ classifies all principal symbol classes of classic pseudodifferential elliptic operators acting by multiplication at infinity and she extended in this context Atiyah-Singer's index theorem, in particular for this case Ind_X computes as well the Fredholm index. Now, if W is a manifold with non compact boundary, we still have the cobordism extension of Carvalho's result via the $K-$theory group $K^1(T_{nc}W)$, however we do not have for the moment the analytic interpretation even if is not very hard to guess it will correspond to b-operators acting by multiplication at the infinity. This will be worked out somewhere else, where we will also use our method to extend the cobordism invariance results to higher and more singular situations.

References

[AS63] M. F. Atiyah and I. M. Singer, *The index of elliptic operators on compact manifolds*, Bull. Amer. Math. Soc. **69** (1963), 422–433. MR0157392

[AS68] M. F. Atiyah and I. M. Singer, *The index of elliptic operators. I*, Ann. of Math. (2) **87** (1968), 484–530. MR0236950

[CR08] P. Carrillo Rouse, *A Schwartz type algebra for the tangent groupoid*, K-theory and noncommutative geometry, EMS Ser. Congr. Rep., Eur. Math. Soc., Zürich, 2008, pp. 181–199. MR2513337

[CRLM14] P. Carrillo Rouse, J. M. Lescure, and B. Monthubert, *A cohomological formula for the Atiyah-Patodi-Singer index on manifolds with boundary*, J. Topol. Anal. **6** (2014), no. 1, 27–74. MR3190137

[Car05] C. Carvalho, *A K-theory proof of the cobordism invariance of the index*, K-Theory **36** (2005), no. 1-2, 1–31 (2006). MR2274155

[Con94] A. Connes, *Noncommutative geometry*, Academic Press, Inc., San Diego, CA, 1994. MR1303779

[Kha13] M. Khalkhali, *Basic noncommutative geometry*, 2nd ed., EMS Series of Lectures in Mathematics, European Mathematical Society (EMS), Zürich, 2013. MR3134494

[Les09] J.-M. Lescure, *Elliptic symbols, elliptic operators and Poincaré duality on conical pseudomanifolds*, J. K-Theory **4** (2009), no. 2, 263–297. MR2551910

[Mon03] B. Monthubert, *Groupoids and pseudodifferential calculus on manifolds with corners*, J. Funct. Anal. **199** (2003), no. 1, 243–286. MR1966830

[MP97] B. Monthubert and F. Pierrot, *Indice analytique et groupoïdes de Lie* (French, with English and French summaries), C. R. Acad. Sci. Paris Sér. I Math. **325** (1997), no. 2, 193–198. MR1467076

[Mor06] S. Moroianu, *On Carvalho's K-theoretic formulation of the cobordism invariance of the index*, Proc. Amer. Math. Soc. **134** (2006), no. 11, 3395–3404. MR2231925

Institut de Mathématiques de Toulouse, 118 route de Narbonne, Toulouse cedex 31062, France

Email address: paulo.carrillo@math.univ-toulouse.fr

Contemporary Mathematics
Volume **709**, 2018
http://dx.doi.org/10.1090/conm/709/14290

Multiplicative ergodic theorems for transfer operators: Towards the identification and analysis of coherent structures in non-autonomous dynamical systems

Cecilia González-Tokman

ABSTRACT. We review state-of-the-art results on multiplicative ergodic theory for operators, with a view towards applications to the analysis of transport phenomena in non-autonomous dynamical systems, such as geophysical flows. The focus of this work is on ideas and motivation, rather than on proofs and technical aspects.

CONTENTS

1. Introduction

An important motivation behind the recent work on multiplicative ergodic theorems is the desire to develop a mathematical theory which is useful for the study of global transport properties of real world dynamical systems, such as oceanic and atmospheric flows.

Global features of the ocean flow include large scale structures which are important for the global climate. One of the best known such structures is the Gulf Stream, which was discovered and navigated by Spanish sailors about five hundred years ago, at the time of the Spanish colonization of the Americas. The Gulf

---

2010 *Mathematics Subject Classification.* Primary 37H15; Secondary 37L55, 47A35, 37N10, 37M25.

Key words and phrases. Multiplicative ergodic theorems, transfer operators, non-autonomous dynamical systems.

This work was supported by ARC DE160100147.

Stream transports water from the tropical latitudes in the warm Caribbean towards the cold North Atlantic, and it is of high importance for the global climate system. Another large scale coherent oceanic structure is the North Pacific Gyre. There, water sinks propitiating conditions for the Great Pacific Garbage Patch to be formed[1]. This patch was discovered only twenty years ago, by Captain Charles Moore. It concentrates plastic residues mostly in the form of microplastics, at the surface and beneath it, and affects marine life in various ways. This fact highlights the importance of developing tools which are capable of analyzing three-dimensional systems.

In the atmosphere, it is well known that there exist vortices near the poles which are persistent over time scales of the order of weeks or longer. The conditions on the Antarctic polar vortex are favorable for chemical reactions responsible for the depletion of ozone, which manifest themselves in the ozone hole. Figure 1 shows streamlines of a snapshot of the wind vector field around the Antarctic polar vortex on September 13th, 2015, from `http://earth.nullschool.net/#current/wind/isobaric/10hPa/orthographic=-70.56,-81.09,288`. Both polar vortices retain cold air, especially in the winter, and change their location over time. This fact illustrates that a time-dependent (or non-autonomous) point of view is necessary for the study of such structures.

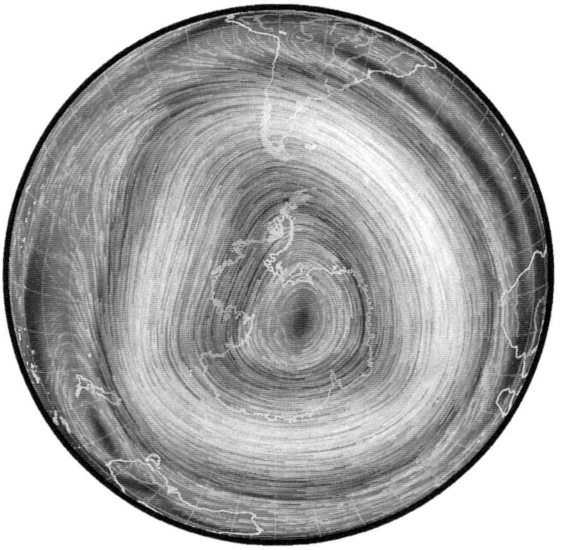

FIGURE 1. Visualization of the Antarctic polar vortex in September 2015. Credit: Cameron Beccario, `http://earth.nullschool.net`. Data from GFS, NCEP, National Weather Service.

[1]The reader may refer to the following National Geographic educational article on this topic: `http://education.nationalgeographic.com/encyclopedia/great-pacific-garbage-patch/`

Two key features of these so-called *coherent structures* which make their analysis challenging, both computationally and mathematically, are: (i) The structures in question are not completely invariant, but rather interact weakly with their surroundings. Hence, especially in the three dimensional setting, it is often unclear how to identify or determine their boundaries; and (ii) The structures are not static, but change their shape and location over time.

A natural framework in which one may attempt to address these challenges is that of *non-autonomous dynamical systems*, introduced in Section 2. Such a setting deals with systems evolving under time-dependent mechanisms, and hence can conveniently account for non static features such as the coherent structures of interest. Non-autonomous dynamical systems have received considerable attention in the literature over the last few years. In fact, some instances thereof –the so-called noisy and driven systems– have been identified as important directions going forward in dynamics in the recent survey article by L.-S. Young [**97**].

One of the tools which has proved to be very successful for analyzing the long term behavior of a large class of dynamical systems is the so-called *transfer operator*, discussed in Section 3. The idea behind this approach is to take a probabilistic point of view of the system, and represent its state at a given time via a density function, describing the distribution of mass (or fluid, or a passive scalar) in the system. The transfer operator is a linear operator that encodes the action of the dynamics on this density. That is, if ρ represents the state of the system at the present time, and $\mathcal{L}$ is the transfer operator associated to the one-step dynamics, then $\mathcal{L}\rho$ describes the state of the system one time step ahead. The crucial factor is that spectral properties (eigenvalues and eigenvectors) of the transfer operator are closely related to objects of dynamical interest.

Section 4 is about *multiplicative ergodic theorems*, dating back to the work of Oseledets [**85**] in the mid 1960s. They provide time-dependent versions of spectral decompositions for composition of random (or non-autonomous) linear operators. Given the considerable success that spectral methods have enjoyed in explaining autonomous dynamical systems via transfer operators, multiplicative ergodic theorems seem to be an ideal candidate for the investigation of coherent structures in the non-autonomous setting. This was the breakthrough idea of Froyland, Lloyd and Quas [**42**], which has been widely investigated in recent years.

The design, development, implementation and optimization of numerical and computational algorithms to identify and visualize relevant features, such as almost invariant sets and coherent structures for models of dynamical systems of physical interest, has paralleled the theoretical development for transfer operators and multiplicative ergodic theorems. This is the topic of Section 5.

Section 6 discusses recent works which have used the technology of transfer operators and multiplicative ergodic theorems to locate coherent structures in models of oceanic and atmospheric flows. A list of online resources for visualizing oceanic and atmospheric flows is presented in Appendix A.

1.1. Acknowledgments. The author thanks her colleagues and collaborators, especially Gary Froyland and Anthony Quas, for years of valuable collaboration and discussions on the topic of the present work; and an anonymous referee for providing valuable comments and references.

This paper was started at the conference "*II Reunión de Matemáticos Mexicanos en el Mundo (MMM2014)*", held at CIMAT, Mexico in December 2014. The author is thankful to the organizers: Octavio Arizmendi, Noé Bárcenas, Fernando Galaz García, José Malagón, Mónica Moreno Rocha, Juan Carlos Pardo Millán, Rodolfo Ríos-Zertuche and Pedro Solórzano, as well as to the sponsors of the event. The author acknowledges support of the Australian Research Council Discovery Early Career Researcher Award DE160100147.

2. Non-autonomous dynamical systems

The difference between autonomous and non-autonomous systems is that the rules dictating the dynamics are fixed in the former, and change over time in the latter. Thus, non-autonomous dynamical systems may be used to describe the evolution of a wider class of phenomena, including systems which are forced or driven, like a planet (or even a group of planets) whose evolution is affected by external forces, such as the presence of one or more other planets, stars and perhaps other factors.

We remark that it is, at least in theory, possible to enlarge a non-autonomous system to make it autonomous, by incorporating all the external factors into it. An important benefit of considering the non-autonomous point of view is that one may focus on the feature of interest – for example, one planet – and obtain much finer conclusions about its behavior than one could get from considering a larger autonomous system, at least with the current technology.

A strategy which has been successful for extending existing theory for autonomous systems to the non-autonomous setting is to consider still objects which explain or illustrate some dynamical feature in the autonomous setting, and animate them into movies, capturing the corresponding characteristic in the non-autonomous setup. An example which has been highly studied is that of a so-called random fixed point: this is simply a point in state space which is moving over time. This approach has been used to describe the asymptotic behavior in some classes of random dynamical systems.

Non-autonomous dynamical systems are also known as random, forced and time-dependent systems, as well as skew-products or cocycles. They consist of a *driving system*: a probability preserving transformation $(\Omega, \mathcal{F}, \mathbb{P}, \sigma)$, and an *evolution rule*. The latter consists of an indexed family of maps $\mathcal{L}_\omega := \mathcal{L}(1, \omega, \cdot)$, where $\mathcal{L} : \mathbb{T} \times \Omega \times X \to X$ is called the *generator* of the system, and is generally assumed to be a measurable function. Here, we will focus on the case of one sided discrete time systems, i.e. $\mathbb{T} = \mathbb{N} := \{0, 1, 2, \dots\}$, but other common choices are $\mathbb{T} \in \{\mathbb{Z}, \mathbb{R}, \mathbb{R}^+\}$, allowing for two sided and continuous time systems as well. Thus, the full non-autonomous dynamical systems is described by the tuple $(\Omega, \mathcal{F}, \mathbb{P}, \sigma, X, \mathcal{L})$. A systematic treatment of these systems is presented in the 1998 book by L. Arnold [**2**].

2.1. The driving system. The driving system encodes the external influences on the system of interest. It is modeled by a probability preserving transformation σ of a probability space $(\Omega, \mathcal{F}, \mathbb{P})$. That is, $\mathbb{P}(\sigma^{-1}E) = \mathbb{P}(E)$ for every measurable set $E \in \mathcal{F}$. We assume σ is invertible[2] and ergodic; that is, if $E \in \mathcal{F}$ is such that $\sigma^{-1}E = E$, then $\mathbb{P}(E) \in \{0, 1\}$.

[2]This feature can often be obtained by considering the natural extension.

This scenario is very flexible. For example, it allows for modeling deterministic driving, such as quasi-periodic forcing, e.g. $\Omega = S^1$, $\mathbb{P} = \text{Leb}$, $\sigma(\omega) = \omega + \alpha \pmod 1), \alpha \notin \mathbb{Q}$. Furthermore, the setting also allows for handling a stochastic framework, in which the external forcing can reflect the influence of noise, for example by setting $\Omega = [-\epsilon, \epsilon]^{\mathbb{Z}}$, $\mathbb{P}$ a product of uniform measures, and σ the shift. In addition, this setting generalizes the autonomous case, which can be recovered by choosing $\Omega = \{\omega_0\}, \mathbb{P} = \delta_{\omega_0}, \sigma = \text{Id}$.

2.2. The evolution rule. For each element $\omega \in \Omega$ there is a corresponding evolution operator $\mathcal{L}_\omega : X \to X$. Here X is the space where the non-autonomous dynamical system evolves. A useful way to think about the whole system is to regard Ω as representing time. Then, $\mathcal{L}_\omega$ corresponds to the evolution rule at the instant ω. The so-called *cocycle property* is necessary to describe the evolution rule for the cocycle after $n \in \mathbb{N}$ steps:

$$\mathcal{L}_\omega^{(n)} := \mathcal{L}_{\sigma^{n-1}\omega} \circ \cdots \circ \mathcal{L}_{\sigma\omega} \circ \mathcal{L}_\omega. \tag{1}$$

When the maps $\mathcal{L}_\omega$ are linear transformations, the cocycle is called a *linear cocycle.*

Among the most extensively studied linear cocycles is the so-called derivative cocycle, which arises from taking a differentiable map $f : M \to M$ as the base transformation, and constructing the cocycle by composing derivatives $\{Df_x\}_{x \in M}$. In this case, $\sigma = f$, $\Omega = M$, $\mathbb{P}$ must be a probability measure preserved by f, $X = T_x M \cong \mathbb{R}^{\dim M}$, and $\mathcal{L}_x = Df_x$. The chain rule implies that the matrix obtained by composition of the matrices along the cocycle exactly describes the linearization (derivative) of the evolution of the system from time 0 to time n: $D(f^n)_x = Df_{f^{n-1}x} \circ \cdots \circ Df_{fx} \circ Df_x$.

In the applications discussed in this work, two other types of evolution operators are considered. In the first one, X is finite, $X = \{B_1, \ldots, B_n\}$. This choice corresponds to a partition of the state space for a non-autonomous dynamical systems of interest, with evolution rules dictated by $T_\omega : M \to M$, where $M = \cup_{i=1}^n B_j$. The evolution rule associated to $\omega \in \Omega$ is in this case a matrix, A_ω, obtained by the so-called Ulam's method [**92**]: $(A_\omega)_{ij} = \frac{m(B_i \cap T_\omega^{-1} B_j)}{m(B_i)}$. In words, the ij-th entry of A_ω is the fraction of orbits starting in B_i which fall inside B_j under one step of the dynamics T_ω, conditional on starting in B_i. This construction relies on having a reference measure m on M, which in applications can frequently be taken to be Lebesgue measure. This is a very intuitive finite-state Markovian model approximating the dynamics of the system.

Another type of evolution operator, highly relevant to the problems at hand, is given by transfer operators. These are bounded linear operators $\mathcal{L}_\omega$ acting on an infinite dimensional Banach space $X = \mathcal{B}$, which encode the evolution of densities or ensembles of particles under the dynamics T_ω. These operators will be the topic of Section 3.

3. Transfer operators and quasi-compactness

Transfer operators yield powerful tools for the study of transport in dynamical systems. Instead of representing the pointwise dynamics of the system, they encode the evolution of densities or ensembles under the dynamics, as illustrated in Figure 2. In many instances, transfer operators may be shown to be bounded linear operators, but typically have the difficulty of acting on infinite dimensional Hilbert or Banach spaces.

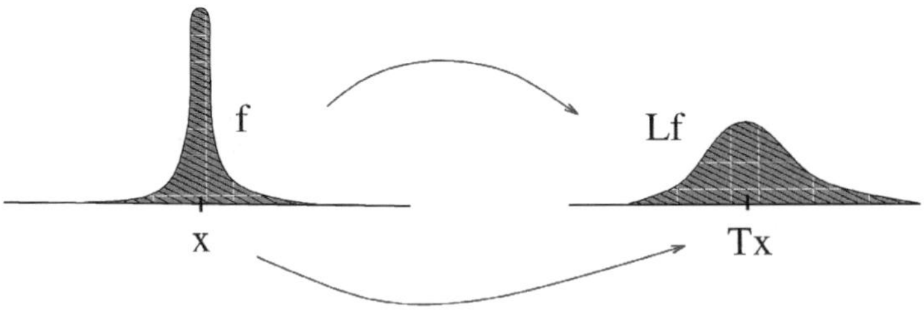

FIGURE 2. Schematic representation of the action of a transfer operator L on a density f, and its relation with the pointwise action of T.

Consider a probability space $(M, \mathcal{M}, m)$ and a non-singular[3] dynamical system $T : M \to M$. The transfer operator L associated to T is a linear operator $L : X \to X$ defined via the duality relation:

$$\int f \cdot g \circ T \, dm = \int Lf \cdot g \, dm, \text{ for every } f \in X, g \in X^*.$$

Here, X can be thought of a space such as $L^1(m)$, but many other possibilities have been used in the literature. In fact, the choice of X is often part of the challenge faced in concrete applications.

3.1. The autonomous setting. A concept which has been key to the current understanding of transport properties of dynamical systems is that of *quasi-compactness.* This property was investigated in the work of Ionescu Tulcea and Marinescu [**59**], and later also by Nussbaum and Hennion [**57, 83**].

A bounded linear operator L acting on a Banach space $(X, \|\cdot\|)$ is called quasi-compact if its spectral radius ρ is strictly larger than its essential spectral radius ρ_e. The *spectral radius* of L is defined as $\rho(L) := \lim_{n\to\infty} \|L^n\|^{1/n}$, and one way of defining the *essential spectral radius* of L is $\rho_e(L) := \lim_{n\to\infty} ic_X(L^n)^{1/n}$, where

$$ic_X(A) := \inf\{r > 0 : A(B_X) \text{ can be covered with finitely many balls of radius } r\},$$

and $B_X = \{x \in X : \|x\| \leq 1\}$ denotes the unit ball of X. Existence of the limit follows in both cases from submultiplicativity. It is straightforward to see that $\rho \geq \rho_e$, and also that L is a compact operator if and only if $\rho_e = 0$.

When L is quasi-compact and $\tau > \rho_e(L)$, the spectrum of L outside the disc of radius τ consists only of finitely many eigenvalues, $\gamma_1, \dots, \gamma_l$ of finite (algebraic) multiplicity. When the leading eigenvalue is simple, the system is said to have a *spectral gap.* As in the case of compact operators, it may be that eigenvalues of L accumulate on the disc of radius $\rho_e(L)$.

Furthermore, if L is quasi-compact, it may be written as follows:

$$L = \sum_{i=1}^{l} (\gamma_i P_i + D_i) + R,$$

[3] $T : M \to M$ is non-singular if $m(T^{-1}(E)) = 0$ for every $E \in \mathcal{M}$ such that $m(E) = 0$.

where P_i is the eigen-projection and D_i the eigen-nilpotent corresponding to γ_i, [**9**, §1.3], [**63**, §III.5]. In addition, $\rho(R) < |\gamma_l|$, and the following relations hold.

$$P_iP_j = \delta_{ij}P_i, \quad P_iD_i = D_iP_i = D_i, \quad (L-\gamma_i)P_i = D_i, \quad P_iR = RP_i = 0.$$

In particular, these imply that

$$L^n = \sum_{i=1}^{l}(\gamma_i P_i + D_i)^n + R^n.$$

Thus, quasi-compactness allows for the spectral analysis of the largest magnitude eigenvalues of L, which are dynamically the most important.

When quasi-compactness holds in the context of transfer operators, the leading eigenvalue is always $\gamma_1 = 1$. The eigenvalues of large modulus and their corresponding (generalized) eigenspaces have been associated with fundamental properties of the underlying dynamical system.

3.1.1. *Peripheral eigenvalues.* The spectral decomposition for the transfer operator (acting on the space of functions of bounded variation, $X = BV$) was used by Hofbauer, Keller [**58**] and Rychlik [**88**] in the study of of piecewise smooth, piecewise expanding maps of the interval. In those works, it was shown that the multiplicity k_1 of the leading eigenvalue 1 provides the number of ergodic components of positive Lebesgue measure – these are considered the relevant ones from a *physical* point of view. A basis for the associated eigenspace may be taken as a set $\{\phi_1, \dots, \phi_k\}$ of non-negative functions with disjoint supports $\pmod m$.

Each of the ϕ_j gives rise to an *absolutely continuous invariant measure (acim)* μ_j defined by $\frac{d\mu_j}{dm} = \phi_j$, describing the long term statistical behavior of $m-$almost every trajectory in the support of ϕ_j, $M_j = supp(\phi_j)$. This is in essentially the same way that the stationary distribution of a finite-state Markov chain describes the long term statistical behavior of almost all realizations, via the law of large numbers: For $m-$almost every $x \in M_j$,

$$\lim_{N\to\infty} \frac{1}{N}\sum_{n=0}^{N-1} f(T^n x) = \int f\phi_j dm.$$

Furthermore, when 1 is simple ($k_1 = 1$), the existence of unit-length eigenvalues different from 1 is the only obstruction for the system to have the so-called mixing property[4]. In fact, if $|\gamma_2| < 1$, then the exponential decay of correlations property holds: Let μ be the measure given by $\frac{d\mu}{dm} = \phi_1$. Then, for every $f \in X, g \in X^*$,

$$\Big|\int f\cdot g\circ T^N d\mu - \int f d\mu \int g d\mu\Big| \le C\|f\|_X \|g\|_{X^*} |\gamma_2|^N.$$

Finer statistical properties, such as central limit theorems, weak invariant principles and laws of iterated logarithms have also been established from spectral properties. The seminal work of Keller and Liverani [**65**] shows stability properties of the spectrum in this and more general settings.

Thanks to the great amount of work devoted to the study of transfer operators in the last decades, these properties are now known to hold in much more general settings than piecewise smooth expanding maps in BV. They essentially rely on

[4]The system is mixing with respect to a measure μ if for every $A, B \in \mathcal{M}$, $\lim_{N\to\infty} \mu(A \cap T^{-N}B) = \mu(A)\mu(B)$.

establishing the quasi-compactness property in an adequate Banach space. Examples include higher dimensional (piecewise) expanding systems, as well as some (piecewise) hyperbolic systems. The book by V. Baladi [**9**] contains references up to the year 2000. More recent work in this direction includes [**10, 12, 15, 27, 56**].

3.1.2. *Sub-unit eigenvalues.* In the 1990's, it was suggested by a group of applied mathematicians, including Dellnitz, Junge, Deuflhard and Schütte, that sub-unit eigenvalues of the transfer operators together with their associated eigenspaces also encompass important dynamical information [**25, 26, 28**]. For example, they can encode the location of so-called metastable regions or almost invariant components.

It remained for some time unknown whether sub-unit isolated eigenvalues could be found in some well studied dynamical systems settings. For instance, the first explicit example of an analytic expanding map of the circle with a sub-unit eigenvalue larger than the essential spectral radius was only found in 2004 [**67**]. Since then, the development of the corresponding theory clarifying the connection between metastability and spectral properties of the transfer operator has received considerable attention [**7, 31, 45, 49, 52, 55, 66, 90**].

This approach has been successfully used by Froyland and collaborators in the study of models of oceanic flow, for example to detect gyres in the Southern ocean [**46**].

3.1.3. *The Koopman operator and a duality relation.* The Koopman or composition operator $\mathcal{K} : X^* \to X^*$, given by $g \mapsto g \circ T$, is the dual of the transfer operator in that $\int Lf \cdot g dm = \int f \cdot g \circ T dm$ for every $f \in X, g \in X^*$.

The Koopman operator is receiving considerable attention in the study of dynamical systems because direct information regarding the evolution of the Koopman operator can be recovered from data measurements [**20**]. For example, suppose T describes the evolution of a fluid over one unit of time, and g is a function measuring temperature, or a similar observable quantity, at different moments in time and along a large number of sensors placed in different spatial locations. Then, each set of measurements taken after j units of time yields information regarding the corresponding power $\mathcal{K}^j = g \circ T^j$.

Despite being closely related, the transfer operator and Koopman operator points of view of the dynamics have so far been considered mostly separately in the literature. An approach which combines the dual perspectives is discussed in Section 4.3.

3.2. The non-autonomous setting.

Two common approaches to the study of non-autonomous systems are the *annealed* and *quenched* points of view. In the annealed case, the focus is on identifying properties of an averaged system which describe global features of the non-autonomous dynamics. This approach has been successful in the study of systems which are driven by a random iid process, such as deterministic systems perturbed by noise [**13, 65**].

The independence property in the driving system allows for the construction of an annealed transfer operator, $\bar{\mathcal{L}} := \int_\Omega \mathcal{L}_\omega d\mathbb{P}(\omega)$, which in some sense represents the behavior of the non-autonomous system. However, when the driving process is not iid, the operator $\bar{\mathcal{L}}$ does not necessarily reflect dynamical properties of the underlying non-autonomous system. In fact, it is not known whether in this more general case there exists an annealed operator reflecting relevant dynamical properties.

The quenched perspective seeks to understand the system by considering individual realizations of the driving process. Early works combining this approach with transfer operators technology include [**8, 11, 21, 22, 68, 69**].

Buzzi [**22**] investigated non-autonomous compositions of piecewise smooth interval maps, under fairly general assumptions on expansion and *distortion*. He established the existence of at least one and at most finitely many so-called *random absolutely continuous invariant measures (acims)*, $\mu = \{\mu_\omega\}_{\omega\in\Omega}$, each with an associated basin $B(\mu) \subset M$ of positive Lebesgue measure, and whose union covers M (mod m). The probability measures μ_ω have associated densities $\phi_\omega = \frac{d\mu_\omega}{dm}$ satisfying the equivariance relation $\mathcal{L}_\omega \phi_\omega = \phi_{\sigma\omega}$. This condition ensures that random acims also satisfy a law of large numbers type condition: For $m-$almost every $x \in B(\mu)$,

$$\lim_{N\to\infty} \frac{1}{N} \sum_{n=0}^{N-1} f(T_\omega^{(n)} x) = \int\int f d\mu_\omega d\mathbb{P}(\omega).$$

In subsequent work [**21**], Buzzi identified conditions ensuring exponential decay of random correlations, as follows:

$$\Big| \int f \cdot g \circ T_\omega^{(N)} d\mu_\omega - \int f d\mu_\omega \int g d\mu_{\sigma^n\omega} \Big| \leq C(\omega) \|f\|_{BV} \|g\|_\infty |\gamma_2|^N.$$

Limit theorems have also been investigated in the non-autonomous setting. Kifer [**70**] introduced a time-dependent centering term, and obtained a version of the central limit theorem, as well as other limit laws. Without the time-dependent centering, this type of limit theorems have only been established in cases where all maps preserve a common invariant measure [**1, 4, 82**].

A key component of the recent progress in the study of transport in non-autonomous systems has been the formulation of a non-autonomous version of the quasi-compactness property, going back to Thieullen [**91**]. A non-autonomous dynamical systems $\mathcal{R} = (\Omega, \mathcal{F}, \mathbb{P}, \sigma, X, \mathcal{L})$ is called quasi-compact if its *maximal Lyapunov exponent*,

$$\Lambda(\mathcal{R}) := \lim_{n\to\infty} \frac{1}{n} \log \|\mathcal{L}_\omega^{(n)}\|_X,$$

is strictly larger than its *index of compactness*,

$$\kappa(\mathcal{R}) := \lim_{n\to\infty} \tfrac{1}{n} \log \mathrm{ic}_X(\mathcal{L}_\omega^{(n)}),$$

where ic_X was defined in Section 3. The fact that these limits exist and are constant $\mathbb{P}-$almost everywhere is a consequence of (i) submultiplicativity of both $\|\cdot\|_X$ and $ic_X(\cdot)$, (ii) ergodicity of the driving system σ and (iii) Kingman's sub-additive ergodic theorem [**71**]. As in the autonomous case, it is straightforward to show that $\Lambda(\mathcal{R}) \geq \kappa(\mathcal{R})$.

The non-autonomous quasi-compactness property has allowed for the possibility of extending the use of transfer operator technology to the study of coherent structures, which are the analogue of metastable regions in the non-autonomous setting. This approach has been facilitated by the so-called multiplicative ergodic theorems, discussed in Section 4.

4. Multiplicative ergodic theorems and coherent structures

Multiplicative ergodic theorems give rise to a time-dependent decomposition which plays the role of a spectral decomposition in the non-autonomous setup.

The so-called *Oseledets splitting* has been used to describe the dominant transport features of the dynamics via non-linear, time-varying modes called *coherent structures*, which decay slowly over time. The rate of decay of each structure is measured by an asymptotic exponential rate, called Lyapunov exponent. This is in contrast with the one-step contraction (or expansion) factor, which an eigenvalue would yield.

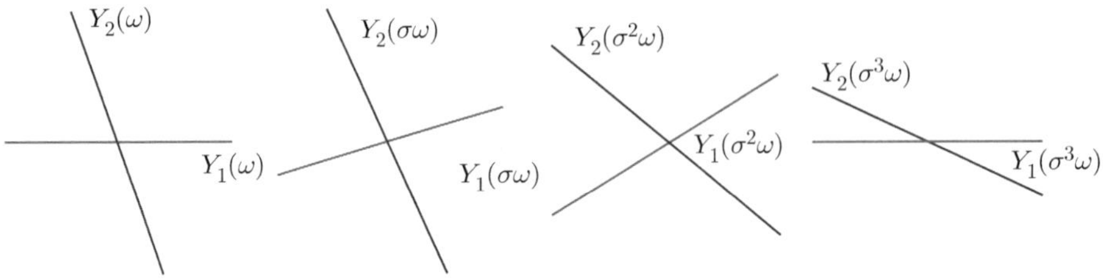

FIGURE 3. Schematic representation of the evolution of Oseledets splittings under a non-autonomous dynamical system.

Intuitively, one may think of each of the components of the splitting as describing one evolving feature of the system, such as a vortex, or an eddy. However, in practice, recovering the actual spatial features from the decomposition requires significantly more work, and there is still considerable research effort taking place in that direction.

The study of multiplicative ergodic theorems covers the cases of finite as well as infinite dimensional underlying spaces X. The finite dimensional case includes settings such as discrete time finite-state Markov chains. The infinite dimensional setting is especially relevant for analyzing more general models of physical phenomena, including the framework of transfer operators. Another distinction among multiplicative ergodic theorems considers whether the evolution rule is invertible or not. While both cases have been investigated in the literature from the start, it is only recently that results guaranteeing a splitting have been established for non-invertible, more precisely semi-invertible, cocycles.

4.1. Definitions and convention. Consider a Banach space X. The *Grassmannian* of X, $\mathcal{G} = \mathcal{G}(X)$, is the set of closed linear subspaces of X, which have closed complements. Included in $\mathcal{G}(X)$ are all finite dimensional and finite codimensional subspaces $\mathcal{G}_k(X) = \{E \subset X, \dim E = k < \infty\}$ and $\mathcal{G}^k(X) = \{F \subset X, \operatorname{codim} F = k < \infty\}$. In particular, when X is a finite dimensional space, $\mathcal{G}(X)$ is the set of all linear subspaces of X.

The Grassmannian $\mathcal{G}(X)$ can be endowed with a metric, defined by $d_{\mathcal{G}}(E, F) := d_H(E \cap B_X, F \cap B_X)$ where B_X denotes the unit ball in X and d_H the Hausdorff distance, $d_H(A, B) := \max\{\sup_{a\in A} \inf_{b\in B} \|a - b\|, \sup_{b\in B} \inf_{a\in A} \|a - b\|\}$. Some fundamental properties of $\mathcal{G}(X)$ are established in [**63**] and, in the context of separable X[5], in [**53**, Appendix B].

[5]A Banach space is called *separable* if it has a countable dense subset.

An *Oseledets splitting* for a linear cocycle $\mathcal{R} = (\Omega, \mathcal{F}, \mathbb{P}, \sigma, X, \mathcal{L})$ consists of:

(I) A sequence of isolated (exceptional) *Lyapunov exponents* $\infty > \lambda_1 > \cdots > \lambda_l > \kappa \geq -\infty$, where the index $l \geq 1$ is allowed to be finite or countably infinite; and

(II) A family of ω-dependent splittings $X = \bigoplus_{j=1}^{l} Y_j(\omega) \oplus V(\omega)$, where $\dim Y_j(\omega) < \infty$ for every $1 \leq j \leq l$, and $V(\omega) \in \mathcal{G}(X)$,

satisfying the following conditions:

(i) Measurability:
The maps $\omega \mapsto Y_j(\omega)$ and $\omega \mapsto V(\omega)$ are $(\mathcal{F}, \mathcal{B}(\mathcal{G}))$ measurable, where $\mathcal{B}(\mathcal{G})$ is the Borel σ-algebra of the Grassmannian $\mathcal{G}(X)$.

(ii) Equivariance:
For each $1 \leq j \leq l$ and $\mathbb{P}$-a.e. $\omega \in \Omega$,

$$\mathcal{L}_\omega Y_j(\omega) = Y_j(\sigma\omega) \text{ and } \mathcal{L}_\omega V(\omega) \subseteq V(\sigma\omega).$$

(iii) Growth rates:
For every $1 \leq j \leq l$ and $\mathbb{P}$-a.e. $\omega \in \Omega$,

$$\lim_{n\to\infty} \tfrac{1}{n} \log \|\mathcal{L}_\omega^{(n)} y\| = \lambda_j \text{ for every } y \in Y_j(\omega) \setminus \{0\}, \text{ and}$$
$$\lim_{n\to\infty} \tfrac{1}{n} \log \|\mathcal{L}_\omega^{(n)} v\| \leq \kappa \text{ for every } v \in V(\omega).$$

(iv) Tempered projections:
Let $\Pi_j(\omega) = \Pi_{Y_j(\omega)\|\oplus_{i\neq j} Y_i(\omega)\oplus V(\omega)}$ and $\Pi'_j(\omega) = \Pi_{\oplus_{i\neq j} Y_i(\omega)\oplus V(\omega)\|Y_j(\omega)}$[6]. Then,

$$\lim_{n\to\pm\infty} \tfrac{1}{n} \log \|\Pi_j(\sigma^n\omega)\| = 0 = \lim_{n\to\pm\infty} \tfrac{1}{n} \log \|\Pi'_j(\sigma^n\omega)\|.$$

It is worth mentioning that, in the context of invertible cocycles, the original work of Oseledets and many subsequent works state, instead of property (iv), a condition on sub-exponential decay of angles between subspaces along trajectories. For instance, for $j' \neq j$,

$$\lim_{n\to\pm\infty} \log \sin \angle(Y_j(\sigma^n\omega), Y_{j'}(\sigma^n\omega)) = 0.$$

Property (iv) implies this alternative statement, when the notion of angle between subspaces, $\angle(E, F)$, (or sine thereof) is extended to Banach spaces, for example as the Grassmannian distance between them, $d_{\mathcal{G}}(E, F)$. This follows from the fact that if $d_{\mathcal{G}}(E, F) < \delta$, then for every $e \in E$ of norm 1, there exists $f \in F$ so that $\|e - f\| < \delta$. In particular, since $\Pi_{E\|F}(e - f) = e$, then $\|\Pi_{E\|F}\| > \delta^{-1}$.

Throughout this section, we will say that the *multiplicative ergodic theorem* holds for the non-autonomous dynamical system $\mathcal{R} = (\Omega, \mathcal{F}, \mathbb{P}, \sigma, X, \mathcal{L})$ if $\mathcal{R}$ has an Oseledets splitting. However, we note that the term is sometimes used in a more general way in the literature.

[6]When the spaces $E, F \in \mathcal{G}(X)$ have trivial intersection, $E \cap F = \{0\}$, the operator $\Pi_{E\|F}$ denotes *projection onto* E *along* F; that is, $\Pi_{E\|F}(e + f) = e$ for every $e \in E, f \in F$.

4.2. Brief history.

4.2.1. *Finite dimensional systems.* Multiplicative ergodic theorems (METs) were first discovered in the mid 1960s by Oseledets [**85**] in the context of finite-dimensional linear cocycles. The existence of an Oseledets splitting was established for non-autonomous dynamical systems of the form $(\Omega, \mathcal{F}, \mathbb{P}, \sigma, \mathbb{R}^n, A)$, where $A : \Omega \to GL(n, \mathbb{R})$ is an invertible matrix valued cocycle satisfying the integrability conditions $\int_\Omega \log^+ \|A(\omega)\| d\mathbb{P}(\omega) < \infty$ and $\int_\Omega \log^+ \|A^{-1}(\omega)\| d\mathbb{P}(\omega) < \infty$.

Furthermore [**85**] addressed the non-invertible case, where $A : \Omega \to M_{n\times n}(\mathbb{R})$ satisfies $\int_\Omega \log^+ \|A(\omega)\| d\mathbb{P}(\omega) < \infty$, and also $\sigma : \Omega \to \Omega$ is allowed to be non-invertible. In this setting, a coarser conclusion was provided, for $\mathbb{P}$-a.e. $\omega \in \Omega$: Aside from the sequence of Lyapunov exponents, there exists a measurable family of ω-dependent *filtrations*, $X = V_1(\omega) \supset V_2(\omega) \cdots \supset V_l(\omega) \supset \{0\}$, which are equivariant in the sense that $A(\omega)V_j(\omega) \subseteq V_j(\sigma\omega)$, and satisfy the growth rate condition $\lim_{n\to\infty} \frac{1}{n} \log \|A^{(n)}(\omega)v\| \leq \lambda_j$, for every $v \in V_j(\omega)$.

In 1979, Raghunathan [**86**] provided an alternative proof of the multiplicative ergodic theorem of Oseledets relying on the singular value decomposition (SVD). This approach has turned out to be useful in much more general settings, and is also at the base of numerical methods for approximation of Oseledets splittings, discussed in Section 5.

One of the major motivations for the development of multiplicative ergodic theorems has been their power in establishing the existence of *stable and unstable manifolds* in dynamical systems. The book by Barreira and Pesin [**14**] includes a detailed treatment of the multiplicative ergodic theorem in the setup of smooth ergodic theory.

Further generalizations of the multiplicative ergodic theorem have been established in other settings, including semi-simple Lie groups [**60**], non-positively curved spaces [**62**], and isometry groups of metric spaces [**61**].

In 2010, Froyland, Lloyd and Quas [**42**] extended Oseledets' theorem to the case of *semi-invertible* cocycles. They established the multiplicative ergodic theorem for finite dimensional linear cocycles $(\Omega, \mathcal{F}, \mathbb{P}, \sigma, \mathbb{R}^n, A)$ under the assumptions of invertibility of σ and integrability of $\log^+ \|A(\omega)\|$, without any invertibility hypotheses on the matrices $A(\omega)$.

4.2.2. *Infinite dimensional systems.* The first multiplicative ergodic theorems in the infinite dimensional case are due to Ruelle [**87**] and Mañé [**76**]. They showed the existence of Oseledets splittings for injective, compact operators acting on separable Hilbert and Banach spaces, respectively. Ruelle treated discrete as well as continuous time systems, and had in mind applications to partial differential equations arising in the study of hydrodynamic turbulence. Mañé treated discrete time systems and aimed for applications to parabolic semilinear equations and delay functional differential equations.

Thieullen [**91**] extended Mañé's result to cocycles of non-autonomous quasi-compact operators on Banach spaces, satisfying a continuity property called $\mathbb{P}$-continuity. In the early 1990s, related results were also established in the context of stochastic partial differential equations [**89**] and stochastic linear delay equations [**77**].

Lian and Lu [**75**] established a version of the multiplicative ergodic theorem which relaxed the continuity condition of [**91**] to a (strong) measurability condition, at the cost of requiring the Banach space X to be separable.

Having in mind applications to the detection of coherent structures in oceanic and atmospheric dynamics, the recent works of Froyland, Lloyd, Quas and González-Tokman [**43**, **53**, **54**] extend infinite dimensional versions of the multiplicative ergodic theorem to the setting of semi-invertible cocycles. Based on [**91**], Froyland, Lloyd and Quas established a multiplicative ergodic theorem in the context of $\mathbb{P}$-continuous cocycles [**43**]. Building on [**75**] and [**30**], Quas and González-Tokman replaced the continuity condition by a strong measurability condition, together with the requirement that the Banach space X (and its dual X^*) is separable [**53**, **54**]. A related volume-based approach to the multiplicative ergodic theorem on Banach spaces is presented in [**16**].

4.3. Identification of Oseledets splittings and coherent structures. The breakthrough paper of Froyland, Lloyd and Quas [**42**] made the connection between the hierarchical structure provided by multiplicative ergodic theorems and coherent structures in non-autonomous dynamical systems. It was for the first time then that ideas such as the second Oseledets space encoding information regarding a dynamical structure which, while changing its location over time, retains a large portion of its *mass* as it evolves were put on firm mathematical grounds. It is also worth remarking that the theoretical advances of [**42**] were followed by a numerical demonstration of their applicability in non-autonomous models of fluid flow in [**44**], a paper in which the term *coherent sets* was introduced.

Next, we present a recent, general method for the identification of Oseledets splittings in the context of linear cocycles on Banach spaces, following [**54**]. Applications of the theory to the identification of coherent structures in geophysical flows are discussed in Section 6. We recall that a singular value decomposition (SVD)[7] of a real $d \times d$ matrix M consists of a diagonal matrix D, and orthogonal matrices U and V such that $M = UDV^T$, where V^T is the transpose of V. The diagonal entries s_j of D are the singular values of M, which will be considered in decreasing order $s_1 \geq s_2 \geq \cdots \geq s_d$. The column vectors of U and V are called left and right singular vectors of M, respectively. If v_j denotes the jth column vector of V, then $Mv_j = s_j u_j$. Hence, the vectors $v_1, \ldots, v_d$ form an orthonormal basis of $\mathbb{R}^d$ which is transformed by M into the orthogonal basis $s_1 u_1, \ldots, s_d u_d$.

It is worth remarking that the case of top singular values and singular vectors of a matrix M are easy to visualize. The top singular value is the norm of M and the associated right singular vector(s) is (are) the unit length vectors v for which the norm of M is realized, $\|Mv\| = \|M\|$. The corresponding left singular vector(s) is (are) the normalized images of the right singular vector(s). Similarly, in the non-autonomous setting, the top Oseledets space and top Lyapunov exponent correspond to the fastest growing vectors and maximal exponential growth rate for the composition of matrices along the cocycle, respectively.

In the finite dimensional setting, the approach of Raghunathan [**86**] gives rise to a method for identifying the Lyapunov exponents and the components of the Oseledets filtration, $\bigoplus_{i \geq j} Y_i(\omega)$, by considering SVDs of the matrices $A^{(n)}(\omega)$. Indeed, the Lyapunov exponents are $\lambda_j = \lim_{n\to\infty} \frac{1}{n} \log s_{d_j}(A^{(n)}(\omega))$, where $s_k(M)$ denotes the kth largest singular value of M, and $d_j = 1 + m_1 + \cdots + m_{j-1}$, with

[7]The SVD of a matrix M can be obtained from the eigen-decomposition of the symmetric matrices MM^T and M^TM. It is not always unique, although the singular values as well as the spaces spanned by singular vectors associated to a fixed singular value are uniquely defined, just as eigenspaces are.

$m_0 = 0$ and for each $i \geq 1$, $m_i = \dim Y_i(\omega)$ is the multiplicity of λ_i. Furthermore, $\bigoplus_{i \geq j} Y_i(\omega) = \lim_{n\to\infty} \bigoplus_{i\geq j} E_i(A^{(n)}(\omega))$ for each j, where $E_j(M)$ denotes the m_j-dimensional space spanned by right singular vectors of M with singular values $s_{d_j}, \dots, s_{d_j+m_j-1}$.

The SVD decomposition is not available in the infinite-dimensional Banach space setup, because it relies on the notion of orthogonality. However, an inductive algorithm to identify Oseledets spaces building on the essential properties of SVDs, which is valid in both finite and infinite dimensional settings, is presented in [**54**]. To overcome the lack of the notion of orthogonality, the algorithm also involves the so-called dual cocycle $\mathcal{R}^* = (\Omega, \mathcal{F}, \mathbb{P}, \sigma^{-1}, X^*, \mathcal{L}^*)$, as well as different notions of volume in Banach spaces. Here X^* is the dual of X, and $\mathcal{L}^*_\omega$ is defined as the adjoint to $\mathcal{L}_{\sigma^{-1}\omega}$, $\mathcal{L}^*_\omega := (\mathcal{L}_{\sigma^{-1}\omega})^*$. In particular, $\mathcal{L}^{*(n)}_{\sigma^n\omega} = (\mathcal{L}^{(n)}_\omega)^*$. While the analysis is significantly more technical than in the finite-dimensional case, Lyapunov exponents can also be obtained from the maximal exponential growth rates of volumes of parallelepipeds of increasing dimensions.

An algorithm to identify Oseledets splittings is as follows. Let $V_1^{(n)}(\omega) = X$, and for each $l > 1$, and $n \in \mathbb{N}$, let $k_l = m_0 + \cdots + m_{l-1}$, and $\theta_1^{(n)}, \dots, \theta_{k_l}^{(n)} \in S_{X^*}$ be unit vectors in X^* such that the volume of the parallelepiped with sides $(\mathcal{L}^{(n)}_\omega)^*\theta_1^{(n)}, \dots, (\mathcal{L}^{(n)}_\omega)^*\theta_{k_l}^{(n)}$ is (nearly) maximal over all possible choices of k_l-tuples in S_{X^*}. Let $V_l^{(n)}(\omega) \subset X$ be the annihilator of the linear span of $(\mathcal{L}^{(n)}_\omega)^*\theta_1^{(n)}, \dots,$ $(\mathcal{L}^{(n)}_\omega)^*\theta_{k_l}^{(n)}$,

$$V_l^{(n)}(\omega) = \langle (\mathcal{L}^{(n)}_\omega)^*\theta_1^{(n)}, \dots, (\mathcal{L}^{(n)}_\omega)^*\theta_{k_l}^{(n)} \rangle^\perp.$$

It is shown in [**54**] that the sequence $V_l^{(n)}(\omega)$ has a limit $V_l(\omega)$ as n tends to infinity. This provides an Oseledets filtration, just like in the finite-dimensional approach of Raghunathan.

Calling $V_l^*(\omega)$ the corresponding Oseledets filtration for the dual cocycle $\mathcal{R}^*$ and letting $W_l(\omega) = (V_l^*(\omega))^\perp$, one obtains a collection of equivariant spaces complementary to the $V_l(\omega)$, $X = V_l(\omega) \oplus W_l(\omega)$. The Oseledets splitting is recovered by letting $Y_l(\omega) = V_l(\omega) \cap W_{l+1}(\omega)$.

It is worth noting that the multiplicative ergodic theorems discussed in Section 4.2 can be applied to Koopman operator cocycles, duals of transfer operator cocycles, whenever the necessary hypotheses hold[8].

The level sets associated to vectors in the dominant Oseledets spaces of transfer operator cocycles (or their finite-time approximations) have been used to locate coherent structures in models of non-autonomous dynamical systems, for example, by thresholding singular vectors at optimal values [**48**]. Thus, the algorithm discussed above may assist in validating existing singular-value decomposition based methods, and perhaps also yield new insights for the identification of coherent structures in applications.

4.4. Stability. A topic of paramount interest for applications is that of stability: How much do the behavior and properties of a dynamical system change under perturbations? The answer to this question is highly relevant for modeling, both analytically and numerically, as well as for accounting for noise and random effects which may affect the evolution of the system.

[8]Quasi-compactness and integrability conditions for $\mathcal{K}$ would follow directly from the analogue conditions for $\mathcal{L}$. However, separability of X^* does not follow from separability of X.

In our setting, the stability question translates into whether Oseledets splittings and Lyapunov exponents are stable under perturbations. One of the difficulties of this problem is that for stability to hold, non-trivial conditions have to be imposed on the non-autonomous dynamical systems. Indeed, without extra hypotheses, there are examples where Lyapunov exponents undergo discontinuous changes under perturbations. This is in contrast with the case of autonomous systems, where spectral stability properties of linear operators are known to hold in very broad generality [**63**].

A topic complementary to stability is bifurcation theory, which investigates qualitatively significant changes which can occur when a system undergoes small perturbations. For a treatment of bifurcations in non-autonomous dynamical systems, we refer the reader to the monograph by Kloeden and Rasmussen [**72**].

4.4.1. *Finite dimensional systems.* The question of stability of Lyapunov exponents for invertible matrix cocycles was treated by Young in [**96**], and later by Ledrappier and Young in [**73**]. They showed stability of Lyapunov exponents when sufficiently random perturbations are considered. In a similar context of invertible matrix cocycles, Ochs showed that stability of Lyapunov exponents is equivalent to stability of the associated Oseledets splitting [**84**]. Recently, Froyland, González-Tokman and Quas [**37**] established a finite-dimensional stochastic stability result for Lyapunov exponents and Oseledets splittings associated to semi-invertible matrix cocycles. Further results concerning stability of Lyapunov exponents in different settings are presented in the recent book by Viana [**94**], and in the forthcoming research monograph by Duarte and Klein [**32**].

Negative stability results for finite-dimensional cocycles include [**3**], which shows that an arbitrarily small L^p perturbation can make all Lyapunov exponents collapse to one point. Another mechanism for instability of Lyapunov exponents has been shown in the context of derivative cocycles associated to area-preserving C^1 diffeomorphisms, by Bochi and Viana [**17**]. The idea behind such a method is attributed to Mañé, and it relies on carefully interchanging vectors corresponding to directions of strongest and weakest expansion on *non-uniformly hyperbolic* cocycles, where there is no uniform lower bound for the angles between Oseledets spaces associated to different Lyapunov exponents.

4.4.2. *Infinite dimensional systems.* In the autonomous case, the work of Keller and Liverani [**65**] guarantees stability of isolated eigenvalues and their associated eigenspaces for quasi-compact operators. This technology may be applied in the setting of random iid perturbations of a non-random map, to obtain results for the annealed (averaged) operators. However, the stability problem remains largely open in the general non-autonomous case.

Most of the stability results available in the quenched setup deal with random perturbations of a fixed (non-random) initial map. Baladi, Kondah and Schmitt [**11**] show stability of the random invariant measure for random perturbations of a smooth expanding map. In a similar setting, Baladi obtains stability results for more general equilibrium states in [**8**].

Bogenschütz [**18**] shows stability of Lyapunov exponents and Oseledets splittings under strong hypotheses on the cocycle and the perturbations. An interesting application of this work addresses the case of stability of the splitting for non-autonomous perturbations of an autonomous system. A recent work by Nakano and Wittsten has employed multiplicative ergodic theorems to investigate small, quenched random perturbations in the context of partially expanding maps on the torus [**81**].

A stability result in the context of transfer operators of random maps has been established by Froyland, González-Tokman and Quas in [**36**]. The main result shows stability of the top Oseledets space –the one corresponding to the random invariant measure– under perturbations including model and numerical approximation errors, as well as noisy perturbations of the underlying maps. A relevant feature of this setting, which makes the analysis more tractable than in the case of second and higher components of the Oseledets splitting, is that the leading Lyapunov exponent is always zero and, in fact, there exists a random fixed point.

5. Numerical and computational aspects

The so-called Ulam's method [**92**], introduced in Section 2.2, has provided an intuitive and effective way of analyzing dynamical systems numerically via transfer operators. Figure 4 shows an approximation of the density of the acim for an interval map relying on Ulam's method (solid line, blue in the electronic version), and on averages taken along a long trajectory (dots, red in the electronic version).

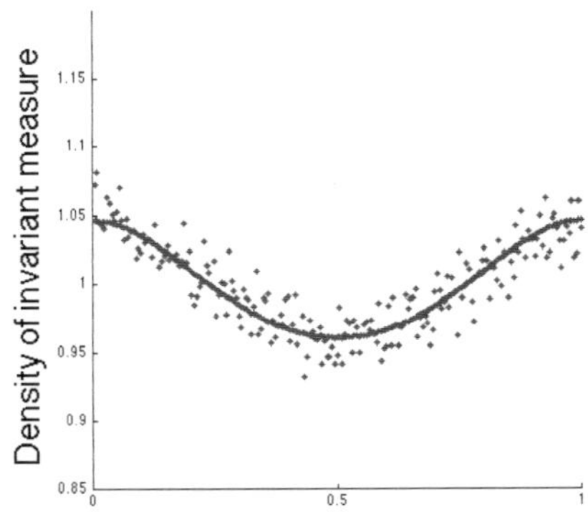

FIGURE 4. Two approximations of the density of the acim of an interval map, obtained via Ulam's method (solid line, blue in the electronic version) and by taking averages over a long trajectory (dots, red in the electronic version).

The rigorous convergence properties of this approximation method, as finer and finer grids are considered, are still open questions in general, even in the autonomous case. This is a particular instance of the stability question discussed in Section 4.4. In some settings, including quite general classes of one-dimensional maps, the scheme has been shown to well approximate acims as well as rates of exponential correlation decay. Some works in this direction are [**5, 6, 19, 29, 33, 64, 65, 74, 78–80**]. Ulam's method was investigated the case of iid and Markovian composition of one-dimensional maps in [**34**].

In the non-autonomous case, efficient numerical algorithms to identify Lyapunov exponents and Oseledets spaces for finite-dimensional (matrix) cocycles have been developed in the last decade by Ginelli and coauthors [**51**], Wolfe and Samelson [**95**] and Froyland, Hüls, Morriss and Watson [**40**]. These approaches rely on very efficient routines built into standard mathematical software, including the QR decomposition and singular-value decomposition algorithms. A comparison of these methods in different model scenarios is presented in [**40**].

Computational algorithms relying on Ulam's method have been used to identify almost invariant sets and coherent structures in autonomous and non-autonomous dynamical systems, for example in the context of geophysical flows, as will be detailed in Section 6. Further numerical methods have been proposed to test for regularity properties of the transfer operator spectrum, with the aim of distinguishing between true and spurious eigenfunctions in [**35**].

The computation of Ulam matrices with high resolution can be a costly algorithm, as it requires to sample all regions of the state space at a high resolution, in fact finer than grid size. On the other hand, once this computation is done, the matrix can be stored for later calculations in the autonomous case. In the non-autonomous setting, a similar procedure is possible, with the additional requirement that, in principle, one has to compute one Ulam matrix for each ω. For these reasons, the quest for more efficient numerical algorithms for the computational analysis of dynamical systems is still a topic of current research efforts [**41, 47**].

6. Applications to oceanic and atmospheric flows

An early application of transfer operator methods to the analysis of geophysical flows was the work of Froyland and collaborators [**46**], where they detect the two major oceanic gyres near Antarctica, the Ross and Wedell seas. The seasonal variability as well as three-dimensional features of these sub polar gyres were subsequently investigated in [**24**].

A related approach has also been used to study "origin, dynamics and evolution of ocean garbage patches from observed surface drifters" in [**93**], and to investigate the global connectedness of the ocean surface in [**50**].

The non-autonomous transfer operator technology has been used to detect and track an eddy of the Agulhas ring, moving around the coast of Africa for several months [**38, 39**]. This analysis was done in two and three dimensions, and showed that such coherent structures are not only displaced horizontally parallel to the surface, but also change along the vertical direction. The outcomes of this approach yield significantly (about 15%) more coherent structures than alternative essentially two-dimensional methods, based on relative vorticity or the so-called Okubo-Weiss parameter.

In the atmosphere, the tools have been used to detect the Antarctic polar vortex and track its evolution over a two-week period [**48**].

A related non-autonomous dynamical systems approach, considering random attractors, has been investigated in stochastic models of climate dynamics in [**23**].

Appendix A. Online resources for the visualization of oceanic and atmospheric flows

Besides the illustrations available in the references provided, there also exist wonderful online resources to visualize oceanic and atmospheric flows. Non-autonomous visualizations of the ocean currents from the years 2005–2007 are available at: `http://svs.gsfc.nasa.gov/cgi-bin/details.cgi?aid=3827`.

An interactive website which contains (autonomous) approximations of ocean currents and wind from recent data is `http://earth.nullschool.net`; see also `http://earth.nullschool.net/about.html` for details. These visualizations are built from data which are constantly being updated. Hence, some features which were visible at the time of writing this work may have changed at the time of reloading the website.

Some of the views available are:

- The Gulf Stream:
`http://earth.nullschool.net/#current/ocean/surface/currents/orthographic=-41.40,29.75,351`
- The Agulhas rings:
`http://earth.nullschool.net/#current/ocean/surface/currents/orthographic=21.28,-18.45,718`
- Oceanic eddies around Australia:
`http://earth.nullschool.net/#current/ocean/surface/currents/orthographic=-222.20,-26.19,1024`
- Winds at surface level:
`http://earth.nullschool.net/#current/wind/surface/level/orthographic=-98.32,31.17,512`
- The Jet stream:
`http://earth.nullschool.net/#current/wind/isobaric/250hPa/equirectangular`
- The Antarctic polar vortex:
`http://earth.nullschool.net/#current/wind/isobaric/10hPa/orthographic=-189.44,-72.33,322`
- The Arctic polar vortex:
`http://earth.nullschool.net/#current/wind/isobaric/10hPa/orthographic=-135.96,59.60,322`

References

[1] R. Aimino, M. Nicol, and S. Vaienti, *Annealed and quenched limit theorems for random expanding dynamical systems*, Probab. Theory Related Fields **162** (2015), no. 1-2, 233–274. MR3350045

[2] L. Arnold, *Random dynamical systems*, Springer Monographs in Mathematics, Springer-Verlag, Berlin, 1998. MR1723992

[3] L. Arnold and N. D. Cong, *Linear cocycles with simple Lyapunov spectrum are dense in* L^∞, Ergodic Theory Dynam. Systems **19** (1999), no. 6, 1389–1404. MR1738943

[4] A. Ayyer, C. Liverani, and M. Stenlund, *Quenched CLT for random toral automorphism*, Discrete Contin. Dyn. Syst. **24** (2009), no. 2, 331–348. MR2486579

[5] W. Bahsoun, *Rigorous numerical approximation of escape rates*, Nonlinearity **19** (2006), no. 11, 2529–2542. MR2267716

[6] W. Bahsoun and C. Bose, *Invariant densities and escape rates: rigorous and computable approximations in the* L^∞*-norm*, Nonlinear Anal. **74** (2011), no. 13, 4481–4495. MR2810743

[7] W. Bahsoun and S. Vaienti, *Metastability of certain intermittent maps*, Nonlinearity **25** (2012), no. 1, 107–124. MR2864379

[8] V. Baladi, *Correlation spectrum of quenched and annealed equilibrium states for random expanding maps*, Comm. Math. Phys. **186** (1997), no. 3, 671–700. MR1463817

[9] V. Baladi, *Positive transfer operators and decay of correlations*, Advanced Series in Nonlinear Dynamics, vol. 16, World Scientific Publishing Co., Inc., River Edge, NJ, 2000. MR1793194

[10] V. Baladi and S. Gouëzel, *Good Banach spaces for piecewise hyperbolic maps via interpolation*, Ann. Inst. H. Poincaré Anal. Non Linéaire **26** (2009), no. 4, 1453–1481. MR2542733

[11] V. Baladi, A. Kondah, and B. Schmitt, *Random correlations for small perturbations of expanding maps*, Random Comput. Dynam. **4** (1996), no. 2-3, 179–204. MR1402416

[12] V. Baladi and M. Tsujii, *Anisotropic Hölder and Sobolev spaces for hyperbolic diffeomorphisms* (English, with English and French summaries), Ann. Inst. Fourier (Grenoble) **57** (2007), no. 1, 127–154. MR2313087

[13] V. Baladi and L.-S. Young, *On the spectra of randomly perturbed expanding maps*, Comm. Math. Phys. **156** (1993), no. 2, 355–385. MR1233850

[14] L. Barreira and Y. B. Pesin, *Lyapunov exponents and smooth ergodic theory*, University Lecture Series, vol. 23, American Mathematical Society, Providence, RI, 2002. MR1862379

[15] M. Blank, G. Keller, and C. Liverani, *Ruelle-Perron-Frobenius spectrum for Anosov maps*, Nonlinearity **15** (2002), no. 6, 1905–1973. MR1938476

[16] A. Blumenthal, *A volume-based approach to the multiplicative ergodic theorem on Banach spaces*, Discrete Contin. Dyn. Syst. **36** (2016), no. 5, 2377–2403. MR3485402

[17] J. Bochi and M. Viana, *The Lyapunov exponents of generic volume-preserving and symplectic maps*, Ann. of Math. (2) **161** (2005), no. 3, 1423–1485. MR2180404

[18] T. Bogenschütz, *Stochastic stability of invariant subspaces*, Ergodic Theory Dynam. Systems **20** (2000), no. 3, 663–680. MR1764921

[19] C. Bose and R. Murray, *The exact rate of approximation in Ulam's method*, Discrete Contin. Dynam. Systems **7** (2001), no. 1, 219–235. MR1806382

[20] M. Budišić, R. Mohr, and I. Mezić, *Applied Koopmanism*, Chaos **22** (2012), no. 4, 047510, 33. MR3388723

[21] J. Buzzi, *Exponential decay of correlations for random Lasota-Yorke maps*, Comm. Math. Phys. **208** (1999), no. 1, 25–54. MR1729876

[22] J. Buzzi, *Absolutely continuous S.R.B. measures for random Lasota-Yorke maps*, Trans. Amer. Math. Soc. **352** (2000), no. 7, 3289–3303. MR1707698

[23] M. D. Chekroun, E. Simonnet, and M. Ghil, *Stochastic climate dynamics: random attractors and time-dependent invariant measures*, Phys. D **240** (2011), no. 21, 1685–1700. MR2844645

[24] M. Dellnitz, G. Froyland, C. Horenkamp, K. Padberg-Gehle, and A. Sen Gupta, *Seasonal variability of the subpolar gyres in the southern ocean: a numerical investigation based on transfer operators*, Nonlinear Processes in Geophysics **16** (2009), no. 6, 655–663.

[25] M. Dellnitz, G. Froyland, and S. Sertl, *On the isolated spectrum of the Perron-Frobenius operator*, Nonlinearity **13** (2000), no. 4, 1171–1188. MR1767953

[26] M. Dellnitz and O. Junge, *On the approximation of complicated dynamical behavior*, SIAM J. Numer. Anal. **36** (1999), no. 2, 491–515. MR1668207

[27] M. F. Demers and C. Liverani, *Stability of statistical properties in two-dimensional piecewise hyperbolic maps*, Trans. Amer. Math. Soc. **360** (2008), no. 9, 4777–4814. MR2403704

[28] P. Deuflhard, W. Huisinga, A. Fischer, and Ch. Schütte, *Identification of almost invariant aggregates in reversible nearly uncoupled Markov chains*, Linear Algebra Appl. **315** (2000), no. 1-3, 39–59. MR1774959

[29] J. Ding and A. Zhou, *Finite approximations of Frobenius-Perron operators. A solution of Ulam's conjecture to multi-dimensional transformations*, Phys. D **92** (1996), no. 1-2, 61–68. MR1384679

[30] T. S. Doan, *Lyapunov exponents for random dynamical systems*, Ph.D. thesis, Fakultät Mathematik und Naturwissenschaften der Technischen Universität Dresden, 2009.

[31] D. Dolgopyat and P. Wright, *The diffusion coefficient for piecewise expanding maps of the interval with metastable states*, Stoch. Dyn. **12** (2012), no. 1, 1150005, 13. MR2887917

[32] P. Duarte and S. Klein, *Lyapunov exponents of linear cocycles*, Atlantis Studies in Dynamical Systems, vol. 3, Atlantis Press, Paris, 2016. Continuity via large deviations. MR3468528

[33] G. Froyland, *Computer-assisted bounds for the rate of decay of correlations*, Comm. Math. Phys. **189** (1997), no. 1, 237–257. MR1478538

[34] G. Froyland, *Ulam's method for random interval maps*, Nonlinearity **12** (1999), no. 4, 1029–1052. MR1709869

[35] G. Froyland, C. González-Tokman, and A. Quas, *Detecting isolated spectrum of transfer and Koopman operators with Fourier analytic tools*, J. Comput. Dyn. **1** (2014), no. 2, 249–278. MR3415255

[36] G. Froyland, C. González-Tokman, and A. Quas, *Stability and approximation of random invariant densities for Lasota–Yorke map cocycles*, Nonlinearity **27** (2014), no. 4, 647.

[37] G. Froyland, C. González-Tokman, and A. Quas, *Stochastic stability of Lyapunov exponents and Oseledets splittings for semi-invertible matrix cocycles*, Comm. Pure Appl. Math. **68** (2015), no. 11, 2052–2081. MR3403759

[38] G. Froyland, C. Horenkamp, V. Rossi, N. Santitissadeekorn, and A. Sen Gupta, *Three-dimensional characterization and tracking of an Agulhas Ring*, Ocean Modelling **52–53** (2012), no. 0, 69 – 75.

[39] G. Froyland, C. Horenkamp, V. Rossi, and E. van Sebille, *Studying an Agulhas ring's long-term pathway and decay with finite-time coherent sets*, Chaos **25** (2015), no. 8, 083119, 9. MR3391356

[40] G. Froyland, T. Hüls, G. P. Morriss, and T. M. Watson, *Computing covariant Lyapunov vectors, Oseledets vectors, and dichotomy projectors: a comparative numerical study*, Phys. D **247** (2013), 18–39. MR3019940

[41] G. Froyland and O. Junge, *On fast computation of finite-time coherent sets using radial basis functions*, Chaos **25** (2015), no. 8, 087409, 11. MR3456020

[42] G. Froyland, S. Lloyd, and A. Quas, *Coherent structures and isolated spectrum for Perron-Frobenius cocycles*, Ergodic Theory Dynam. Systems **30** (2010), no. 3, 729–756. MR2643709

[43] G. Froyland, S. Lloyd, and A. Quas, *A semi-invertible Oseledets theorem with applications to transfer operator cocycles*, Discrete Contin. Dyn. Syst. **33** (2013), no. 9, 3835–3860. MR3038042

[44] G. Froyland, S. Lloyd, and N. Santitissadeekorn, *Coherent sets for nonautonomous dynamical systems*, Phys. D **239** (2010), no. 16, 1527–1541. MR2659535

[45] G. Froyland, R. Murray, and O. Stancevic, *Spectral degeneracy and escape dynamics for intermittent maps with a hole*, Nonlinearity **24** (2011), no. 9, 2435–2463. MR2819930

[46] G. Froyland, K. Padberg, M. England, and A.-M. Treguier, *Detection of coherent oceanic structures via transfer operators*, Phys. Rev. Lett. **98** (2007), no. 22, 224503.

[47] G. Froyland and K. Padberg-Gehle, *A rough-and-ready cluster-based approach for extracting finite-time coherent sets from sparse and incomplete trajectory data*, Chaos **25** (2015), no. 8, 087406, 15. MR3456017

[48] G. Froyland, N. Santitissadeekorn, and A. Monahan, *Transport in time-dependent dynamical systems: finite-time coherent sets*, Chaos **20** (2010), no. 4, 043116, 10. MR2791360

[49] G. Froyland and O. Stancevic, *Escape rates and Perron-Frobenius operators: open and closed dynamical systems*, Discrete Contin. Dyn. Syst. Ser. B **14** (2010), no. 2, 457–472. MR2660868

[50] G. Froyland, R. M. Stuart, and E. van Sebille, *How well-connected is the surface of the global ocean?*, Chaos **24** (2014), no. 3, 033126, 10. MR3404424

[51] F. Ginelli, P. Poggi, A. Turchi, H. Chaté, R. Livi, and A. Politi, *Characterizing dynamics with covariant Lyapunov vectors*, Physical Review Letters **99** (2007), no. 13, 130601.

[52] C. González-Tokman, B. R. Hunt, and P. Wright, *Approximating invariant densities of metastable systems*, Ergodic Theory Dynam. Systems **31** (2011), no. 5, 1345–1361. MR2832249

[53] C. González-Tokman and A. Quas, *A semi-invertible operator Oseledets theorem*, Ergodic Theory Dynam. Systems **34** (2014), no. 4, 1230–1272. MR3227155

[54] C. González-Tokman and A. Quas, *A concise proof of the multiplicative ergodic theorem on Banach spaces*, J. Mod. Dyn. **9** (2015), 237–255. MR3400385

[55] P. Góra, A. Boyarsky, and P. Eslami, *Metastable systems as random maps*, Internat. J. Bifur. Chaos Appl. Sci. Engrg. **22** (2012), no. 11, 1250279, 11. MR3006350

[56] S. Gouëzel and C. Liverani, *Banach spaces adapted to Anosov systems*, Ergodic Theory Dynam. Systems **26** (2006), no. 1, 189–217. MR2201945

[57] H. Hennion, *Sur un théorème spectral et son application aux noyaux lipchitziens* (French, with English summary), Proc. Amer. Math. Soc. **118** (1993), no. 2, 627–634. MR1129880

[58] F. Hofbauer and G. Keller, *Ergodic properties of invariant measures for piecewise monotonic transformations*, Math. Z. **180** (1982), no. 1, 119–140. MR656227

[59] C. T. Ionescu Tulcea and G. Marinescu, *Théorie ergodique pour des classes d'opérations non complètement continues* (French), Ann. of Math. (2) **52** (1950), 140–147. MR0037469

[60] V. A. Kaĭmanovich, *Lyapunov exponents, symmetric spaces and a multiplicative ergodic theorem for semisimple Lie groups* (Russian, with English summary), Zap. Nauchn. Sem. Leningrad. Otdel. Mat. Inst. Steklov. (LOMI) **164** (1987), no. Differentsial′naya Geom. Gruppy Li i Mekh. IX, 29–46, 196–197; English transl., J. Soviet Math. **47** (1989), no. 2, 2387–2398. MR947327

[61] A. Karlsson and F. Ledrappier, *On laws of large numbers for random walks*, Ann. Probab. **34** (2006), no. 5, 1693–1706. MR2271477

[62] A. Karlsson and G. A. Margulis, *A multiplicative ergodic theorem and nonpositively curved spaces*, Comm. Math. Phys. **208** (1999), no. 1, 107–123. MR1729880

[63] T. Kato, *Perturbation theory for linear operators*, Classics in Mathematics, Springer-Verlag, Berlin, 1995. Reprint of the 1980 edition. MR1335452

[64] M. Keane, R. Murray, and L.-S. Young, *Computing invariant measures for expanding circle maps*, Nonlinearity **11** (1998), no. 1, 27–46. MR1492949

[65] G. Keller and C. Liverani, *Stability of the spectrum for transfer operators*, Ann. Scuola Norm. Sup. Pisa Cl. Sci. (4) **28** (1999), no. 1, 141–152. MR1679080

[66] G. Keller and C. Liverani, *Rare events, escape rates and quasistationarity: some exact formulae*, J. Stat. Phys. **135** (2009), no. 3, 519–534. MR2535206

[67] G. Keller and H. H. Rugh, *Eigenfunctions for smooth expanding circle maps*, Nonlinearity **17** (2004), no. 5, 1723–1730. MR2086147

[68] K. Khanin and Y. Kifer, *Thermodynamic formalism for random transformations and statistical mechanics*, Sinaĭ's Moscow Seminar on Dynamical Systems, Amer. Math. Soc. Transl. Ser. 2, vol. 171, Amer. Math. Soc., Providence, RI, 1996, pp. 107–140. MR1359097

[69] Y. Kifer, *Equilibrium states for random expanding transformations*, Random Comput. Dynam. **1** (1992/93), no. 1, 1–31. MR1181378

[70] Y. Kifer, *Limit theorems for random transformations and processes in random environments*, Trans. Amer. Math. Soc. **350** (1998), no. 4, 1481–1518. MR1451607

[71] J. F. C. Kingman, *The ergodic theory of subadditive stochastic processes*, J. Roy. Statist. Soc. Ser. B **30** (1968), 499–510. MR0254907

[72] P. E. Kloeden and M. Rasmussen, *Nonautonomous dynamical systems*, Mathematical Surveys and Monographs, vol. 176, American Mathematical Society, Providence, RI, 2011. MR2808288

[73] F. Ledrappier and L.-S. Young, *Stability of Lyapunov exponents*, Ergodic Theory Dynam. Systems **11** (1991), no. 3, 469–484. MR1125884

[74] T. Y. Li, *Finite approximation for the Frobenius-Perron operator. A solution to Ulam's conjecture*, J. Approximation Theory **17** (1976), no. 2, 177–186. MR0412689

[75] Z. Lian and K. Lu, *Lyapunov exponents and invariant manifolds for random dynamical systems in a Banach space*, Mem. Amer. Math. Soc. **206** (2010), no. 967, vi+106. MR2674952

[76] R. Mañé, *Lyapounov exponents and stable manifolds for compact transformations*, Geometric dynamics (Rio de Janeiro, 1981), Lecture Notes in Math., vol. 1007, Springer, Berlin, 1983, pp. 522–577. MR730286

[77] S. E. A. Mohammed, *The Lyapunov spectrum and stable manifolds for stochastic linear delay equations*, Stochastics Stochastics Rep. **29** (1990), no. 1, 89–131.

[78] R. Murray, *Discrete approximation of invariant densities*, Ph.D. thesis, University of Cambridge, 1997.

[79] R. Murray, *Existence, mixing and approximation of invariant densities for expanding maps on* $\mathbb{R}^r$, Nonlinear Anal. **45** (2001), no. 1, Ser. A: Theory Methods, 37–72. MR1828066

[80] R. Murray, *Ulam's method for some non-uniformly expanding maps*, Discrete Contin. Dyn. Syst. **26** (2010), no. 3, 1007–1018. MR2600727

[81] Y. Nakano and J. Wittsten, *On the spectra of quenched random perturbations of partially expanding maps on the torus*, Nonlinearity **28** (2015), no. 4, 951–1002. MR3320041

[82] P. Nándori, D. Szász, and T. Varjú, *A central limit theorem for time-dependent dynamical systems*, J. Stat. Phys. **146** (2012), no. 6, 1213–1220. MR2903045

[83] R. D. Nussbaum, *The radius of the essential spectrum*, Duke Math. J. **37** (1970), 473–478. MR0264434

[84] G. Ochs, *Stability of Oseledets spaces is equivalent to stability of Lyapunov exponents*, Dynam. Stability Systems **14** (1999), no. 2, 183–201. MR1698361

[85] V. I. Oseledec, *A multiplicative ergodic theorem. Characteristic Ljapunov, exponents of dynamical systems* (Russian), Trudy Moskov. Mat. Obšč. **19** (1968), 179–210. MR0240280

[86] M. S. Raghunathan, *A proof of Oseledec's multiplicative ergodic theorem*, Israel J. Math. **32** (1979), no. 4, 356–362. MR571089

[87] D. Ruelle, *Characteristic exponents and invariant manifolds in Hilbert space*, Ann. of Math. (2) **115** (1982), no. 2, 243–290. MR647807

[88] M. Rychlik, *Bounded variation and invariant measures*, Studia Math. **76** (1983), no. 1, 69–80. MR728198

[89] K.-U. Schaumlöffel and F. Flandoli, *A multiplicative ergodic theorem with applications to a first order stochastic hyperbolic equation in a bounded domain*, Stochastics Stochastics Rep. **34** (1991), no. 3-4, 241–255. MR1124837

[90] C. Schütte and M. Sarich, *Metastability and Markov state models in molecular dynamics*, Courant Lecture Notes in Mathematics, vol. 24, Courant Institute of Mathematical Sciences, New York; American Mathematical Society, Providence, RI, 2013. Modeling, analysis, algorithmic approaches. MR3155191

[91] P. Thieullen, *Fibrés dynamiques asymptotiquement compacts. Exposants de Lyapounov. Entropie. Dimension* (French, with English summary), Ann. Inst. H. Poincaré Anal. Non Linéaire **4** (1987), no. 1, 49–97. MR877991

[92] S. M. Ulam, *A collection of mathematical problems*, Interscience Tracts in Pure and Applied Mathematics, no. 8, Interscience Publishers, New York-London, 1960. MR0120127

[93] E. van Sebille, M.H. England, and G. Froyland, *Origin, dynamics and evolution of ocean garbage patches from observed surface drifters*, Environ. Res. Lett. **7** (2012), no. 4, 044040.

[94] M. Viana, *Lectures on Lyapunov exponents*, Cambridge Studies in Advanced Mathematics, vol. 145, Cambridge University Press, Cambridge, 2014. MR3289050

[95] C.L. Wolfe and R.M. Samelson, *An efficient method for recovering Lyapunov vectors from singular vectors*, Tellus A **59** (2007), no. 3, 355–366.

[96] L.-S. Young, *Random perturbations of matrix cocycles*, Ergodic Theory Dynam. Systems **6** (1986), no. 4, 627–637. MR873436

[97] L.-S. Young, *Understanding chaotic dynamical systems*, Comm. Pure Appl. Math. **66** (2013), no. 9, 1439–1463. MR3078694

School of Mathematics and Physics, The University of Queensland, St Lucia, QLD 4072, Australia

Email address: cecilia.gt@uq.edu.au

Contemporary Mathematics
Volume **709**, 2018
http://dx.doi.org/10.1090/conm/709/14291

Representation stability for the pure cactus group

Rita Jiménez Rolland and Joaquín Maya Duque

ABSTRACT. The fundamental group of the real locus of the Deligne-Mumford compactification of the moduli space of rational curves with n marked points, the pure cactus group, resembles the pure braid group in many ways. As it is the case for several "pure braid like" groups, it is known that its cohomology ring is generated by its first cohomology. In this note we survey what the FI-module theory developed by Church, Ellenberg and Farb can tell us about those examples. As a consequence we obtain uniform representation stability for the sequence of cohomology groups of the pure cactus group.

1. Introduction

In this paper we survey the principal notions and consequences of the FI-module theory introduced by Church, Ellenberg and Farb in [**CEF15**]. Our main objective is to show how the theory can be readily applied to certain sequences of groups and spaces with cohomology rings that have the structure of a graded FI-algebra and are known to be generated by the first cohomology group. We revisit the collection of examples in Section 4 that have in common this behavior (see Table 1).

To illustrate our discussion we work out the details for the sequence of spaces $\{M_n\}$. The space $M_n := \overline{\mathcal{M}}_{0,n}(\mathbb{R})$ is the real locus of the Deligne-Mumford compactification $\overline{\mathcal{M}}_{0,n}$ of the moduli space of rational curves with n marked points. We recall the precise definition of M_n in Section 2 below.

As pointed out by Etingof–Henriques–Kamnitzer–Rains in [**EHKR10**], the space M_n resembles in many ways the configuration space of $n-1$ distinct points in the plane $\mathcal{F}(\mathbb{C}, n-1)$. For instance, the spaces M_n are smooth manifolds and Eilenberg-Mac Lane spaces ([**DJS03**]). In particular, it follows that the cohomology of the *pure cactus groups* $\pi_1(M_n)$ coincides with the cohomology of M_n. Another similarity, there exist maps from M_n to M_{n-1} which can be described as "forgetting one marked point" and from M_n to M_{n+1} which can be thought of "inserting a bubble at a marked point". These maps endow the collection of spaces $\{M_n\}$ with the structure of a simplicial space. Furthermore, as we recall in 2.2, the rational

2010 *Mathematics Subject Classification.* Primary 20J06; Secondary 20C15, 20F36.

Key words and phrases. Pure cactus group, braid groups, moduli space, cohomology, representation stability, FI-modules.

The second author is grateful for the support of the CONACYT grant 168093-F.

cohomology of M_n is generated multiplicatively by the 1-dimensional classes coming from the simplest of such spaces, namely, $M_4 = S^1$.

On the other hand, there are also notable differences between the spaces M_n and $\mathcal{F}(\mathbb{C}, n-1)$. For instance, for $n > 5$ the space M_n is not formal [**EHKR10**]. In contrast to the configuration spaces, the forgetful maps $M_n \to M_{n-1}$ are not fibrations. It is not known whether the pure cactus group $\pi_1(M_n)$ is residually nilpotent.

In the present note we show that the spaces M_n share one more property with the configuration spaces and the examples considered in Section 4. Namely, we see in Section 3 that their rational cohomology ring has the structure of a graded FI-algebra which is a finitely generated FI-module in each degree. As main consequence of this approach we obtain, as in the case of configuration spaces, uniform representation stability in the sense of [**CF13**].

THEOREM 1.1. *The sequence $\{H^i(M_n, \mathbb{Q})\}_n$ of S_n-representations is uniformly representation stable for all $i \geq 0$ and the stability holds for $n \geq 6i$.*

Moreover, we show that the characters of the S_n-representations are strongly constrained.

THEOREM 1.2. *For $i \geq 0$ and $n \geq 6i$ the character of the S_n-representation $H^i(M_n, \mathbb{Q})$ is given by a unique character polynomial P_i of degree $\leq 3i$.*

A *character polynomial* is a polynomial in the cycle-counting functions X_l, where $X_l(\sigma)$ counts the number of l-cycles in the cycle decomposition of $\sigma \in \bigsqcup_n S_n$. For instance, for $n \geq 4$ it is known that $H^1(M_n, \mathbb{Q}) = \bigwedge^3 \mathcal{H}_n$, where $\mathcal{H}_n$ is the standard representation of S_n of dimension $n-1$. Its character is given by the character polynomial of degree 3:

$$\chi_{H^1(M_n,\mathbb{Q})} = \binom{X_1}{3} + X_3 - X_2X_1 - \binom{X_1}{2} + X_2 + X_1 - 1.$$

Notice that for $n \geq 4$, the value of the character in any permutation $\sigma \in S_n$ only depends on cycles of length 1, 2 and 3 in the cycle decomposition of σ. More generally, Theorem 1.2 implies that the character of the S_n-representation $H^i(M_n, \mathbb{Q})$ thought of as a function of $\sigma \in S_n$ is completely independent to the number of r-cycles for all $r > 3i$.

Furthermore, we recover the following result first proven in [**EHKR10**, Theorem 6.4].

COROLLARY 1.3. For every $i \geq 1$, there exists a unique polynomial $p_i \in \mathbb{Q}[t]$ of degree $\leq 3i$ such that for $n \geq 6i$, the ith Betti number of M_n is given by $b_i(M_n) = p_i(n)$.

In particular, for $n \geq 4$ the first Betti number $b_1(M_n)$ is the cubic polynomial in n:

$$\dim_{\mathbb{Q}}\left(H^1(M_n, \mathbb{Q})\right) = \chi_{H^1(M_n,\mathbb{Q})}(\mathrm{id}) = \binom{n}{3} - \binom{n}{2} + n - 1 = \frac{(n-1)(n-2)(n-3)}{6}.$$

More is actually known about these representations. In [**Rai09**, Theorem 1.1] Rains obtained an explicit formula for the graded character of the S_n-action on $H^*(M_n, \mathbb{Q})$. In particular, his result recovers the product formula [**EHKR10**, Theorem 6.4] for the Poincaré series of M_n.

Acknowledgments. The authors would like to thank Benson Farb, Jacob Mostovoy, Bruno Cisneros and Jennifer Wilson for useful conversations and comments. We are grateful to the referee for all the suggestions that helped improve the exposition in this paper. The first author would also like to thank the Centro de Ciencias Matemáticas, UNAM-Morelia, for all its support during the writing of this paper.

2. The space M_n and its cohomology ring

The Deligne-Mumford compactification $\overline{\mathcal{M}}_{0,n}$ of the moduli space $\mathcal{M}_{0,n}$ of rational curves with n marked points is a smooth projective variety over $\mathbb{Q}$ that parametrizes stable curves of genus 0 with n labeled points (see [**DM69**]).

DEFINITION 2.1. A *stable rational curve* with n marked points is a finite union C of projective lines $C_1, C_2, \ldots, C_k$ together with distinct marked points $z_1, \ldots, z_n$ in C such that

- each marked point z_i is in only one C_j;
- $C_i \cap C_j$ is either empty or consists only of one point, and in this case the intersection is transversal;
- its associated graph, with a vertex for each C_i and an edge for each pair of interesection lines, is a tree;
- each component C_i has at least three special points, where a special point is either an intersection point with another component, or a marked point.

In $\overline{\mathcal{M}}_{0,n}$ we say that two stable curves $C = \{C_1, \ldots, C_k, (z_1, \ldots, z_n)\}$ and $C' = \{C'_1, \ldots, C'_k, (z'_1, \ldots, z'_n)\}$ are equivalent if there is an isomorphism of algebraic curves $f : C \to C'$ such that $f(z_i) = z'_i$. Notice f is given by a collection of k fractional linear maps $f_j : C_j \to C'_{\sigma(j)}$, where $\sigma \in S_k$. In particular, if f is an automorphism of C, each f_i is an automorphism of $\mathbb{P}^1$ that fixes at least three points (the special points of the component C_i). Hence a stable curve C has no non-trivial automorphisms.

Let $n, m \geq 3$ and consider an injection $f : [m] \hookrightarrow [n]$, of the set $[m] := \{1, \ldots, m\}$ into $[n] := \{1, \ldots, n\}$. There is a well-defined "forgetful morphism"

$$\phi_f : \overline{\mathcal{M}}_{0,n} \to \overline{\mathcal{M}}_{0,m}$$

given by $\phi_f(C) = \phi_f(\{C_1, \ldots, C_k, (z_1, \ldots, z_n)\}) := \{C'_1, \ldots, C'_r, (z_{f(1)}, \ldots, z_{f(m)})\}$ where the components C'_i of $\phi_f(C)$ are the same as the components C_i of C unless they become unstable (see figure below). Notice that ϕ_f forgets all the marked points whose index are not in the image $f([m])$. If after this "forgetting process" the component C_i has less than three special points, then we contract the component C_i so that $\phi_f(C)$ is still a stable curve (for more details see for example [**KV07**, Chapter 1] and references therein). Furthermore, by taking $f \in$End$([n])$, we obtain a natural action of the symmetric group S_n on $\overline{\mathcal{M}}_{0,n}$ by permuting the marked points.

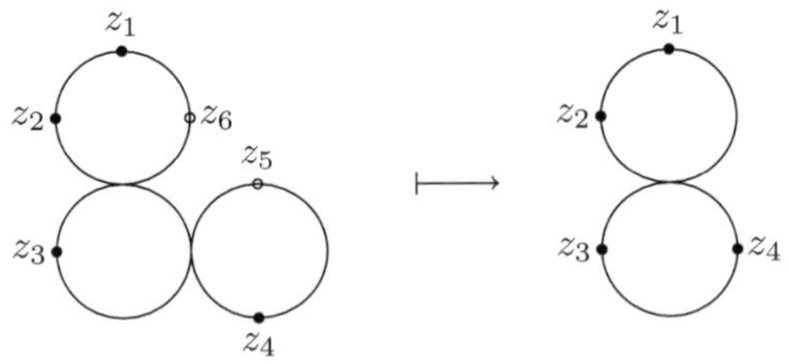

FIGURE 1. Forgetful morphism for the canonical inclusion $[4] \hookrightarrow [6]$.

Our space of interest is the real locus of this variety $M_n := \overline{\mathcal{M}}_{0,n}(\mathbb{R})$. It can be described as the set of equivalence classes of stable curves with n marked points defined over $\mathbb{R}$. Since the projective lines are circles over the real numbers, a stable curve is a tree of circles with labeled points on them (as referred to in [**EHKR10**] as a "cactus-like" structure). The space M_n is a compact and connected smooth manifold of dimension $n-3$ (see [**DJS03**] and [**Dev99**]). It is clear that M_3 is a point and the cross ratio gives us an isomorphism between M_4 and the circle. It can also be shown that M_5 is a connected sum of five real projective planes (see [**Dev99**]).

The cohomology ring of M_n was completely determined in [**EHKR10**]. We recall their description next.

THEOREM 2.2 ([**EHKR10**] 2.9). *Let Λ_n be the skew-commutative algebra generated by the elements of degree one w_{ijkl}, $1 \leq i,j,k,l \leq n$, which are antisymmetric in i,j,k,l*[1] *with defining relations*

$$w_{ijkl} + w_{jklm} + w_{klmi} + w_{lmij} + w_{mijk} = 0$$

$$w_{ijkl}w_{ijkm} = 0.$$

Then Λ_n is isomorphic to $H^*(M_n, \mathbb{Q})$, and the action of S_n is given by

$$\pi(w_{ijkl}) = w_{\pi(i)\pi(j)\pi(k)\pi(l)}.$$

As we mentioned before, for $n, m \geq 3$ and each m-subset S of $[n]$, there is a well-defined forgetful map $\phi_S : M_n \to M_m$ which induces a map in cohomology $\phi_S^* : H^*(M_m, \mathbb{Q}) \to H^*(M_n, \mathbb{Q})$. If $m = 4$ and $S = \{i,j,k,l\}$ with $1 \leq i < j < k < l \leq n$, the induced map $\phi_S^* : H^*(M_4, \mathbb{Q}) \to H^*(M_n, \mathbb{Q})$ takes the generator of $H^*(M_4, \mathbb{Q}) \cong \mathbb{Q}$ to w_{ijkl}.

Observe that since $w_{ijkl} = w_{1jkl} - w_{1kli} + w_{1lij} - w_{1ijk}$, the set

$$\{w_{1ijk} | 1 < i < j < k \leq n\}$$

is a generating set for the algebra Λ_n. Also note that since there are not other linear relations that set is also a basis for the vector space $H^1(M_n, \mathbb{Q})$, we will use this fact below.

[1]Meaning $-w_{ijkl} = w_{jikl} = w_{ikjl} = w_{ijlk}$.

3. Representation stability and FI-modules

In this section we revisit what it means for a sequence of S_n-representations to be uniformily representation stable in the sense of [**CF13**] and we survey the main notions from the theory of FI-modules developed by Church, Ellenberg and Farb in [**CEF15**] that can be used to prove such phenomenon. Our main purpose is to discuss how these ideas can be readily applied to certain sequences of groups or spaces (see examples in Section 4) with cohomology rings that are known to be generated by the first cohomology group. This approach was first introduced in [**CEF15**, Section 4.2] for a more general setting and was applied to another collection of examples in [**CEF15**, Section 5]. We illustrate the concepts with the example given by the cohomology of the moduli spaces M_n and obtain the proof of 1.1.

3.1. Representation stability. Church and Farb introduced in [**CF13**] a notion of stability for certain sequences of S_n- representations over a field k of characteristic zero. We recall their definitions here.

DEFINITION 3.1. A sequence $\{V_n, \phi_n\}_n$ of S_n-representations over k together with homomorphisms $\phi_n : V_n \to V_{n+1}$ is said to be *consistent* if the maps ϕ_n are S_n-equivariant with respect to the natural inclusion $S_n \hookrightarrow S_{n+1}$.

Notation: Over a field of characteristic zero, irreducible representations of S_n are defined over $\mathbb{Q}$ and every S_n-representation decomposes as a direct sum of irreducible representations. Furthermore, irreducible representations of S_n are classified by partitions of n. By a partition of n we mean a collection $\lambda = (l_1, \dots, l_r)$ with $|\lambda| := l_1 + \cdots + l_r = n$ and $l_i \geq l_{i+1} > 0$. We will denote such a partition by $\lambda \vdash n$ and the associated irreducible representation by V_λ. Following the notation in [**CEF15**], given a partition $\lambda = (l_1, \dots, l_j) \vdash m$ of a positive integer m and an integer $n \geq m + l_1$, we define the S_n representation $V(\lambda)_n$ as

$$V(\lambda)_n := V_{\lambda[n]},$$

the irreducible representation that corresponds to the padded partition $\lambda[n] := (n - m, l_1, \dots, l_j)$ of n.

DEFINITION 3.2. (Uniform representation stability) Let $\{V_n, \phi_n\}_n$ be a consistent sequence of S_n-representations. This sequence is *uniformly representation stable* if there exists a natural number N such that for $n \geq N$ the following conditions are satisfied:

(1) The map ϕ_n is injective.
(2) The span of the S_{n+1}-orbit of $\phi_n(V_n)$ is equal to V_{n+1}.
(3) If V_n is decomposed into irreducible representations as $\bigoplus_\lambda c_{\lambda,n} V(\lambda)_n$, the multiplicities $c_{\lambda,n}$ for each λ are independent of n.

3.2. The FI-module structure. It was noticed in [**CEF15**] that certain consistent sequences of S_n-representations could be encoded as a single object and this perspective had strong advantages. Using their notation, we consider the category FI whose objects are all finite sets and whose morphisms are all injections.

DEFINITION 3.3. An *FI-module* over a commutative ring R is a functor from the category FI to the category of R-modules. If V is an *FI-module* we denote by V_n or $V(\mathbf{n})$ the R-module associated to $\mathbf{n} := [n]$ and by $f_* : V_m \to V_n$ the map corresponding to the inclusion $f \in \mathrm{Hom}_{FI}(\mathbf{m}, \mathbf{n})$.

The category of FI-modules over R is closed under covariant functorial constructions on R-modules, if we apply functors pointwise. In particular, if V and W are FI-modules, then $V \otimes W$ and $V \oplus W$ are FI-modules.

Notice that since $\mathrm{End}_{FI}(\mathbf{n}) = S_n$, an FI-module V encodes the information of the consistent sequence $\{V_n, (I_n)_*\}$ of S_n-representations with the maps induced by the natural inclusion $I_n : [n] \hookrightarrow [n+1]$. It should be noted that not every consistent sequence can be encoded in this way [**CEF15**, Remark 3.3.1].

The FI-modules $H^i(M_\bullet, \mathbb{Q})$. In this paper we are interested in certain sequences of spaces or groups that can be encoded as a contravariant functor $Y_\bullet$ from the category FI to the category of spaces, a *co-FI-space*, or to the category of groups, a *co-FI-group*. Therefore, by composing with the contravariant functors $H^i(-, R)$ and $H^*(-, R)$, we obtain the FI-modules $H^i(Y_\bullet, R)$ over R for each $i \geq 0$ and $H^*(Y_\bullet, R)$ which is a functor from FI to the category of graded R-algebras, what [**CEF15**] calls a *graded FI-algebra* over R, .

In particular, we can think of the sequence of manifolds $\{M_n\}$ as the objects of the co-FI-space $M_\bullet$ that takes each $\mathbf{n}$ to M_n and each inclusion $f \in \mathrm{Hom}_{FI}(\mathbf{m}, \mathbf{n})$ to the corresponding forgetful map $\phi_f : M_n \to M_m$ defined as before. In this section, for each $i \geq 0$ we sometimes denote the FI-module $H^i(M_\bullet, \mathbb{Q})$ simply by H^i and the graded FI-algebra $H^*(M_\bullet, \mathbb{Q})$ by H^*.

3.3. Finite generation of an FI-module.

We are not interested in all FI-modules, but in those that are finitely generated in the following sense.

DEFINITION 3.4. Let V be an FI-module over R.

a) If Σ is a subset of the disjoint union $\bigsqcup_n V_n$ we define $\mathrm{span}_V(\Sigma)$, the *span of* Σ, to be the minimal sub-FI-module of V containing Σ and we say that Σ *generates* the FI-module $\mathrm{span}_V(\Sigma)$.
b) We say that V is *generated in degree* $\leq m$ if $\mathrm{span}_V\left(\bigsqcup_{k\leq m} V_k\right) = V$.
c) We say that V is *finitely generated* if there is a finite set of elements $\{v_1, \ldots, v_k\}$ with $v_i \in V_{n_i}$ which generates V, that is, $\mathrm{span}_V(\{v_1, \ldots, v_k\}) = V$.

Finite generation of FI-modules is a property that is closed under quotients and extensions. It also passes to sub-FI-modules when R is a field that contains $\mathbb{Q}$ (see [**CEF15**, Theorem 1.3]) and more generally when R is a Noetherian ring (see [**CEFN14**, Theorem A]). It is key that finite generation is closed under tensor products and we have control of the degree of generation.

PROPOSITION 3.5. [**CEF15**, Prop. 2.3.6] *If V and W are finitely generated FI-modules, so is $V \otimes W$. If V is generated in degree $\leq m_1$ and W is generated in degree $\leq m_2$, then $V \otimes W$ is generated in degree $\leq m_1 + m_2$.*

Our examples below are graded FI-algebras and we will take advantage of this structure to obtain finite generation.

DEFINITION 3.6. Let A be a graded FI-algebra over R. Given a graded sub-FI-module V of A, we say that A *is generated by* V if V_n generates A_n as an R-algebra for all $n \geq 0$.

From 2.2 , the graded FI-algebra $H^*(M_\bullet, R)$ is generated by the FI-module $H^1(M_\bullet, R)$ (which can be thought as the graded sub-FI-module concentrated in grading 1). In all the examples in Section 4, the graded FI-algebra of interest $H^*(Y_\bullet, R)$ is a quotient of the free tensor algebra on the generating classes in

$H^1(Y_\bullet, R)$. Therefore, Proposition 3.5 reduces the question of finite generation for the FI-modules $H^i(Y_\bullet, R)$ to a question of finite generation for the FI-module $H^1(Y_\bullet, R)$.

PROPOSITION 3.7. Suppose that the FI-module $H^1(Y_\bullet, R)$ is finitely generated in degree $\leq g$. If the graded FI-algebra $H^*(Y_\bullet, R)$ is generated by $H^1(Y_\bullet, R)$, then for each $i \geq 1$ the FI-module $H^i(Y_\bullet, R)$ is finitely generated in degree $\leq g \cdot i$.

Degree of generation of $H^i(M_\bullet, \mathbb{Q})$. Using the explicit description of the cohomology ring in 2.2 we can obtain finite generation for the FI-modules H^i from Proposition 3.7. Moreover, we can obtain the following upper bound for the degree of generation.

Notice that, in particular, H^1 is finitely generated in degree ≤ 4 and Proposition 3.7 only implies that H^i is finitely generated in degree $\leq 4i$.

LEMMA 3.8. *The FI-module H^i is finitely generated in degree $\leq 3i + 1$.*

PROOF. Recall that $H^1(M_n, \mathbb{Q})$ is generated by elements w_{1klm} with k, l, m between 2 and n, and also that it generates the whole cohomology ring. Hence the vector space $H^i(M_n, \mathbb{Q})$ is generated by elements

$$x = w_{1k_1l_1m_1} \cdots w_{1k_il_im_i}$$

with the indices varying from 2 to n. A generator element x has at most $3i + 1$ different indices in the set $[n]$. Then, after the action of some permutation σ of S_n, $\sigma \cdot x$ is in the image of $H^i(M_{3i+1}, \mathbb{Q})$ inside $H^i(M_n, \mathbb{Q})$. Hence

$$x = \sigma^{-1}(\sigma \cdot x) \in \operatorname{span}_{H^i}(H^i_{3i+1}),$$

therefore $H^i = \operatorname{span}_{H^i}(H^i_{3i+1})$. Since $H^i(M_{3i+1}, \mathbb{Q})$ is of finite dimension, we conclude that the FI-module H^i is finitely generated in degree $\leq 3i + 1$. □

One of the highlights of the theory of FI-modules is that finite generation of an FI-module over a field k of characteristic zero is equivalent to uniform representation stability of the corresponding consistent sequence.

THEOREM 3.9. [**CEF15**, Thm 1.13 and Prop. 3.3.3] *An FI-module over k is finitely generated if and only if the sequence $\{V_n, (I_n)_*\}$ of S_n-representations is uniformly representation stable and each V_n is finite dimensional. Furthermore, the stable range is $n \geq s + d$ where d and s are the weight and the stability degree of V respectively.*

In Sections 3.4 and 3.5 below we recall the definitions of weight and stability degree of an FI-module. We focus on FI-modules over a field k of characteristic zero.

Representation stability for the cohomology of M_n. We just proved finite generation for the FI-module H^i in Lemma 3.8. Therefore, the theorem above implies uniform representation stability for the sequence $\{H^i(M_n, \mathbb{Q})\}$. The specific stable range in 1.1 follows from the computations of weight and stability degree of the FI-module H^i in Lemmas 3.15 and 3.19 below.

REMARK 3.10. Representation stability for the cohomology of the pure cactus groups can also be obtained by the methods in [**CF13**] where the notion was first introduced. This approach, however, allows us to prove only the stability of $H^i(M_n, \mathbb{Q})$ with respect to the S_{n-1}-action (rather than the action of S_n) and we obtain a worse estimate for the stable range than the one achieved using FI-modules.

3.4. Weight of an FI-module. It turns out that finite generation of an FI-module V puts certain constraints on the partitions (shape of the Young diagrams) in the irreducible representations of each representation V_n as we discuss below.

DEFINITION 3.11. Let V be an FI-module over k. We say that V has *weight* d if, for all $n \geq 0$, d is the maximum order $|\lambda|$ over all the irreducible constituents $V(\lambda)_n$ of V_n. We write weight$(V) = d$.

By definition, if W is a subquotient of V , then weight$(W) \leq$ weight(V). Moreover, the degree of generation of an FI-module gives an upper bound for the weight.

PROPOSITION 3.12. [**CEF15**, Prop. 3.2.5] If the FI-module V over k is generated in degree $\leq g$, then weight$(V) \leq g$.

This implies, for example, that for a finitely generated FI-module V the alternating representation $V(\lambda)_n$, where $\lambda = (1, \ldots, 1)$ has $|\lambda| = n-1$, cannot appear in the decomposition into irreducibles for $n \gg 0$. We also have control of the weight under tensor products.

PROPOSITION 3.13. [**CEF15**, Prop. 3.2.2] If V and W are FI-modules over k, then weight$(V \otimes W) \leq$ weight(V)+weight(W).

As a consequence we have the following result that can be applied to our examples of interest.

COROLLARY 3.14. If the graded FI-algebra $H^*(Y_\bullet, k)$ is generated by the FI-module $H^1(Y_\bullet, k)$ and weight$\big(H^1(Y_\bullet, k)\big) \leq d$, then for each $i \geq 1$ we have that weight$\big(H^i(Y_\bullet, k)\big) \leq d \cdot i$.

The weight of $H^i(M_\bullet, \mathbb{Q})$. From Proposition 3.12 and our computation above we have that the weight of $H^i(M_\bullet, \mathbb{Q})$ is bounded above by $3i + 1$. We can do slightly better.

Let $\mathcal{H}_n$ be the standard representation of S_n, the subrepresentation of dimension $n-1$ of the permutation representation $\mathbb{Q}^n$ consisting of vectors with coordinates that add to zero. We consider the map $H^1(M_n, \mathbb{Q}) \to \bigwedge^3 \mathcal{H}_n$ given by $w_{ijkl} \mapsto (e_i - e_l) \wedge (e_j - e_l) \wedge (e_k - e_l)$ and observe that it respects the anti-symmetry and the five terms relation from the description in 2.2. Moreover, the map is clearly surjective and a dimension count shows that it is an isomorphism of S_n-representations as pointed out in [**EHKR10**, Prop. 2.2].

The S_n-representation $\bigwedge^3 \mathcal{H}_n$ is irreducible and corresponds to the partition $(n-3, 1, 1, 1)$ of n. With our notation above this means that $H^1(M_n, \mathbb{Q}) = V(1, 1, 1)$ for $n \geq 4$ and weight$\big(H^1(M_\bullet, \mathbb{Q})\big) = 3$. From Proposition 3.14 we obtain that weight$\big(H^i(M_\bullet, \mathbb{Q})\big) \leq 3i$. Hence we have shown.

LEMMA 3.15. *The FI-module H^i has weight $\leq 3i$.*

3.5. Stability degree of an FI-module. In order to define the stability degree we recall the notion of coinvariants of a representation.

Let V_n be an S_n-representation. The coinvariants $(V_n)_{S_n}$ is the largest S_n-equivariant quotient of V_n on which S_n acts trivially. Equivalently, it is the module $V_n \otimes_{k[S_n]} k$. If V is an FI-module, we can apply this construction to all the V_n simultaneously, obtaining the k-modules $(V_n)_{S_n}$ for each $n \geq 0$. Moreover, all the maps $V_n \to V_{n+1}$ involved in the definition of the FI-module structure induce a single map $T : (V_n)_{S_n} \to (V_{n+1})_{S_{n+1}}$.

For each integer $a \geq 0$ fix once and for all some set $\bar{\mathbf{a}}$ of cardinality a, for instance, take $\bar{\mathbf{a}} = \{-1, \dots, -a\}$. We consider the action of S_n in $V_{\bar{\mathbf{a}} \sqcup \mathbf{n}}$ induced by the elements in $\mathrm{End}_{FI}(\bar{\mathbf{a}} \sqcup \mathbf{n})$ that are the identity on $\bar{\mathbf{a}}$ and take the coinvariants $(V_{\bar{\mathbf{a}} \sqcup \mathbf{n}})_{S_n}$.

DEFINITION 3.16. The *stability degree* stab-deg(V) of an FI-module V is the smallest $s \geq 0$ such that for all $a \geq 0$, the map $T : (V_{\bar{\mathbf{a}} \sqcup \mathbf{n}})_{S_n} \to (V_{\bar{\mathbf{a}} \sqcup \mathbf{n+1}})_{S_{n+1}}$, induced by any inclusion $[n] \hookrightarrow [n+1]$, is an isomorphism for all $n \geq s$.

For FI-modules over $\mathbb{Q}$, when taking tensor products, the stability degree is still bounded above.

PROPOSITION 3.17. [**KM18**, Prop. 2.23] If the FI-modules V and W over $\mathbb{Q}$ have stability degree $\leq s_1, s_2$ and weight $\leq d_1, d_2$ respectively, then $V \otimes W$ has stability degree $\leq \max\{s_1 + d_1, s_2 + d_2, d_1 + d_2\}$.

This implies the following result in the setting of interest for this paper.

COROLLARY 3.18. If the graded FI-algebra $H^*(Y_\bullet, \mathbb{Q})$ is generated by the FI-module $H^1(Y_\bullet, \mathbb{Q})$ with weight$\leq d$ and stability degree $\leq s$, then for each $i \geq 1$ we have that stab-deg$\big(H^i(Y_\bullet, \mathbb{Q})\big) \leq \max\{s + d, d \cdot i\}$.

Stability degree of $H^i(M_\bullet, \mathbb{Q})$. We now find an upper bound for the stability degree for our example of interest.

LEMMA 3.19. *The FI-module H^i has stability degree $\leq 3i$.*

PROOF. First of all we describe the vector space $H^1(M_{|\bar{\mathbf{a}} \sqcup \mathbf{n}|}, \mathbb{Q})$ and the action of S_n on it. The generating elements of this vector space are of the form w_{ijkl} where the indexes are now in the set $\bar{\mathbf{a}} \sqcup [n]$ and the only linear relations among them are the usual five term relations, hence if $n \geq 1$ the set $\{w_{1jkl} | j, k, l \in \bar{\mathbf{a}} \sqcup [n]\}$ is a generating set for this vector space and S_n acts only in the indexes in $[n]$. Given $a \geq 0$, denote by

$$(V_{\bar{\mathbf{a}} \sqcup \mathbf{n}})_n = \big(H^1(M_{|\bar{\mathbf{a}} \sqcup \mathbf{n}|}, \mathbb{Q})\big)_{S_n}$$

the vector space of coinvariants of $H^1(M_{|\bar{\mathbf{a}} \sqcup \mathbf{n}|}, \mathbb{Q})$ where S_n acts only in the indices in $[n] = \{1, \dots, n\}$. Therefore $(V_{\bar{\mathbf{a}} \sqcup \mathbf{n}})_n = H^1(M_{|\bar{\mathbf{a}} \sqcup \mathbf{n}|}, \mathbb{Q})/W$, where W is the subspace generated by $\{v - \sigma \cdot v |\ v \in H^1(M_{|\bar{\mathbf{a}} \sqcup \mathbf{n}|}, \mathbb{Q}) \text{ and } \sigma \in S_n\}$. We claim that $B = \{w_{1jkl} | \text{ at least one } j, k, l \in [n]\}$ is a generating set for W. First of all notice that if $j \in [n]$ we have

$$w_{1jkl} = \frac{1}{2}(w_{1jkl} - (1\ j) \cdot w_{1jkl})$$

and $w_{1jkl} \in W$, similarly if k or l belongs to $[n]$. On the other hand, note that if j, k, l are not in $[n]$ and $\sigma \in S_n$ we have

$$\sigma \cdot w_{1jkl} = w_{\sigma(1)jkl} = w_{1jkl} - w_{kl1\sigma(1)} - w_{l1\sigma(1)j} - w_{1\sigma(1)jk},$$

then the difference $w_{1jkl} - \sigma \cdot w_{1jkl}$ is in the span of B. Moreover, notice that if $w_{1jkl} \in B$, its image by any permutation is in W. Therefore B spans W and it is a basis, up to the anti-symmetry of the indexes, for W. Hence, for $n \geq 1$, $(V_{\bar{\mathbf{a}}\sqcup\mathbf{n}})_n$ has $\{w_{1jkl} | \ j < k < l \text{ and } j,k,l \in \bar{\mathbf{a}}\}$ as basis. Observe that this basis does not depend on n. Since the map

$$T : (V_{\bar{\mathbf{a}}\sqcup\mathbf{n}})_n \to (V_{\bar{\mathbf{a}}\sqcup\mathbf{n+1}})_{n+1}$$

takes the basis to itself, it is an isomorphism for $n \geq 1$.

Therefore stab-deg(H^1)≤ 1. If $|\bar{\mathbf{a}}| = 3$ we have $(V_{\bar{\mathbf{a}}\sqcup\mathbf{0}})_0 = H^1(M_3, \mathbb{Q}) = 0$ and $(V_{\bar{\mathbf{a}}\sqcup\mathbf{1}})_1 = H^1(M_4, \mathbb{Q}) = \mathbb{Q}$ thus stab-deg(H^1)$\neq 0$ therefore stab-deg(H^1)=1. Since we know that weight(H^1) $= 3$, the upper bound for the stability degree of H^i follows from Corollary 3.18.

□

3.6. Polynomiality of characters. We end this section by discussing another way that finite generation of FI-modules imposes strong constraints in the corresponding S_n-representations. If V is finitely generated, then the sequence of characters χ_{V_n} of the S_n-representations V_n is *eventually polynomial* as we now explain. For each $i \geq 1$, we consider the class function $X_l : \bigsqcup_n S_n \to \mathbb{Z}$ defined by

$$X_l(\sigma) = \text{ number of } l\text{-cycles in the cycle decomposition of } \sigma.$$

As mentioned in the Introduction, polynomials in the variables X_l are called *character polynomials*. The degree of the character polynomial $P(X_1, \ldots X_r)$ is defined by setting $\deg(X_l) = l$.

THEOREM 3.20 ([**CEF15**]Theorem 3.3.4). *Let V be an FI-module over $\mathbb{Q}$. If V is finitely generated then there exist a unique character polynomial $P_V \in \mathbb{Q}[X_1, X_2, \ldots, X_d]$ with $\deg P_V \leq d$ such that for all $n \geq s + d$*

$$\chi_{V_n}(\sigma) = P_V(\sigma) \text{ for all } \sigma \in S_n,$$

where d and s are the weight and the stability degree of V respectively.

The upper bound on the degree of the character polynomial P_V in 3.20 implies that P_V only involves the variables $X_1, \ldots, X_d$. This tells us that the character χ_{V_n} only depends on "short cycles"', i.e. on cycles of length $\leq d$, independently of how large n is. Moreover, we have that the dimension of V_n

$$\dim V_n = \chi_{V_n}(\text{id}) = P_V(n, 0, \ldots, 0)$$

is a polynomial in n of degree $\leq d$ for $n \geq s + d$.

Characters for the cohomology of M_n. 3.20 together with the computations below of upper bounds for the weight and the stability degree for the FI-modules H^i give us Theorem 1.2. Corollary 1.3 follows from this theorem and the discussion before.

4. Other examples

We would like to end this note by revisiting the collection of "pure braid like" examples in Table 1 that have been considered using the perspective discussed in Section 3. In general, the bounds in Table 1 are not sharp.

TABLE 1. Examples of "pure braid like" goups and spaces that have cohomology rings with an FI-algebra structure generated by the first cohomology.

Group or Space Y_n	**Cohomology Ring** $H^*(Y_n, \mathbb{Q})$	$H^1(Y_n, \mathbb{Q})$	$\chi_{H^1(Y_n,\mathbb{Q})}$	weight $H^p(Y_\bullet) \leq$	deg-gen $H^p(Y_\bullet) \leq$	stab-deg $H^p(Y_\bullet) \leq$
$\mathcal{F}(\mathbb{C}, n)$	Exterior algebra $\mathcal{R}_n$ generated by $w_{i,j}$, $1 \leq i \neq j \leq n$, with relations $w_{i,j} = w_{j,i}$, $w_{i,j}w_{j,k} + w_{j,k}w_{k,i} + w_{k,i}w_{i,j} = 0$ [**Arn69**]	$V(\cdot) \oplus V(1) \oplus V(2)$ for $n \geq 4$	$\binom{X_1}{2} + X_2$ for $n \geq 0$	$2p$	$2p$	$2p$ [**CEF15**]
$\mathcal{M}_{0,n+1}$	Subalgebra of $\mathcal{R}_n$ generated by 1 and $\theta_{i,j} := w_{i,j} - w_{1,2}$ with $\{i, j\} \neq \{1, 2\}$ [**Gai96**, Cor. 3.1] *& references therein*	$V(1) \oplus V(2)$ for $n \geq 4$	$\binom{X_1}{2} + X_2 - 1$ for $n \geq 2$	$2p$	$4p$	$2p$ [**JR16**]
$\overline{\mathcal{M}}_{0,n}(\mathbb{R})$	Supercommutative algebra generated by w_{ijkl} $1 \leq i, j, k, l \leq n$, antisymmetric in i, j, k, l with relations $w_{ijkl}w_{ijkm} = 0$ and $w_{ijkl} + w_{jklm} + w_{klmi} + w_{lmij} + w_{mijk} = 0$ [**EHKR10**, Thm. 2.9]	$V(1, 1, 1)$ for $n \geq 4$	$\binom{X_1}{3} + X_3 - X_2X_1 - \binom{X_1}{2} + X_2 + X_1 - 1$ for $n \geq 3$	$3p$ (3.15)	$3p + 1$ (3.8)	$3p$ (3.19)
PvB_n	Exterior algebra generated by $w_{i,j}$, $1 \leq i \neq j \leq n$, with relations $w_{i,j}w_{j,i} = 0$, $w_{i,j}w_{i,k} = w_{i,j}w_{j,k} - w_{i,k}w_{k,j}$ and $w_{i,k}w_{j,k} = w_{i,j}w_{j,k} - w_{j,i}w_{i,k}$ [**Lee13**, Section 5] *&* [**BEER06**]	$V(\cdot) \oplus V(1)^2 \oplus V(1, 1) \oplus V(2)$ for $n \geq 4$	$2\binom{X_1}{2}$ for $n \geq 0$	$2p$ [**Lee**]	$2p$	$2p$
PfB_n	Exterior algebra generated by $w_{i,j}$ $1 \leq i \neq j \leq n$, with relations $w_{i,j} = -w_{j,i}$, $w_{i,j}w_{i,k} = w_{i,j}w_{j,k}$ and $w_{i,k}w_{j,k} = w_{i,j}w_{j,k}$ for $i < j < k$ [**Lee13**, Section 5] *&* [**BEER06**]	$V(1) \oplus V(1, 1)$ for $n \geq 3$	$\binom{X_1}{2} - X_2$ for $n \geq 0$	$2p$ [**Lee**]	$2p$	$2p$
$P\Sigma_n$	Exterior algebra generated by $w_{i,j}$ $1 \leq i \neq j \leq n$, with relations $w_{i,j}w_{j,i} = 0$, and $w_{k,j}w_{j,i} - w_{k,j}w_{k,i} + w_{i,j}w_{k,i} = 0$ [**JMM06**]	$V(\cdot) \oplus V(1)^2 \oplus V(1, 1) \oplus V(2)$ for $n \geq 4$	$2\binom{X_1}{2}$ for $n \geq 0$	$2p$ [**Wil15**]	$2p$	$2p$

Configuration space of n points in $\mathbb{C}$ and the pure braid group. This example was the first studied with this approach in [**CEF15**, Example 5.1.A]. We have the co-FI-space $\mathcal{F}(\mathbb{C}, \bullet)$ given by $\mathbf{n} \mapsto \mathcal{F}(\mathbb{C}, n)$. Since these configuration spaces are Eilenberg-MacLane spaces, we could alternatively consider their fundamental groups P_n, the *pure braid groups.* The cohomology ring in this case is the so called Arnol'd algebra which is generated by cohomology classes of degree one. We recall its description in Table 1 and recover the information about the weight and degree of generation from our discussion in Section 3. Uniform representation stability was originally proved in [**CF13**].

Moduli space of n-pointed curves of genus zero. The *moduli space of $(n+1)$-pointed curves of genus zero* $\mathcal{M}_{0,n+1}$ has a natural action of S_{n+1} by permuting the marked points. If we focus on the action of the subgroup S_n generated by the transpositions $(2\ 3), (3\ 4), \ldots, (n\ n+1)$ in S_{n+1}, we can identify its cohomology ring with a subalgebra of $H^*(\mathcal{F}(\mathbb{C}, n), \mathbb{Q})$ as we recall in Table 1. Hence the graded FI-algebra $H^*(\mathcal{M}_{0,\bullet+1}, \mathbb{Q})$ is generated by degree one classes and our arguments in Section 3 readily apply. As explained in [**JR16**], we can use conclusions on these "shifted" FI-modules to obtain information about $H^*(\mathcal{M}_{0,\bullet}, \mathbb{Q})$.

REMARK 4.1. The theory of FI-modules can also be used to study and obtain representation stability of the cohomology of configuration spaces of n points on connected manifolds and moduli spaces of genus $g \geq 2$ with n marked points (see [**Chu12**], [**CEF15**], [**JR11**] and [**JR15**]). In these cases, little information is known about the corresponding cohomology rings and the arguments needed are more involved than the ones discussed in this note.

Pure virtual and flat braid groups. The *pure virtual braid groups* PvB_n and the *pure flat braid groups* PfB_n are generalizations of the braid group P_n that allow *virtual* crossings of strands. Those virtual crossings were introduced in [**Kau99**]. The original motivation was to study knots in thickened surfaces of higher genus and the extension of knot theory to the domain of Gauss codes and Gauss diagrams. Explicit definitions and presentations of these groups can be found in [**Lee**, Section 2.1]. As for the pure braid group P_n, there is a well defined action of S_n on $H^*(PvB_n; \mathbb{Q})$ and $H^*(PfB_n; \mathbb{Q})$. Moreover, we have well-defined forgetful maps compatible with the S_n-action. Hence we can consider the FI-modules $H^p(PvB_\bullet; \mathbb{Q})$ and $H^p(PfB_\bullet; \mathbb{Q})$ in the setting from Section 3.

From the description in Table 1, we see that the corresponding cohomology rings have a graded FI-algebra structure and are generated by the first cohomology group. Moreover, from the explicit decompositions of $H^p(PvB_n, \mathbb{Q})$ and $H^p(PfB_n, \mathbb{Q})$ as a direct sum of induced representations obtained in [**Lee**, Theorem 3 and 8], it follows that $H^p(PvB_\bullet, \mathbb{Q})$ and $H^p(PfB_\bullet, \mathbb{Q})$ have the extra-structure of an $FI\#$-module (see [**CEF15**, Def. 4.1.1 and Thm. 4.1.5]). This in particular implies that the corresponding stability degree is bounded above by the weight ([**CEF15**, Cor. 4.1.8]). Theorem 3.9 implies uniform representation stability, which was previously obtained in [**Lee**] using the approach in [**CF13**].

In [**Wil14**], Wilson extended the work of Church, Ellenberg, Farb, and Nagpal on sequences of representations of S_n to representations of the signed permutation groups B_n and the even-signed permutation groups D_n. She introduced the notion of a finitely generated $FI_{\mathcal{W}}$-module, where $\mathcal{W}_n$ is the Weyl group in type A_{n-1},

B_n/C_n, or D_n. Arguments like the ones discussed in Section 3 also apply in this setting ([**Wil14**, Prop. 5.11]). Uniform representation stability and polynomiality of characters are also consequences of this approach ([**Wil14**, Thms 4.26 & 4.27] and [**Wil15**, Thm 4.16]).

Group of pure string motions. The *group* Σ_n *of string motions* is another generalization of the braid group. It is defined as the group of motions of n smoothly embedded, oriented, unlinked, unknotted circles in $\mathbb{R}^3$ and can be identified with the *symmetric automorphism group* of the free group F_n. We refer the interested reader to [**Wil15**, Section 5.1] and references therein. The subgroup $P\Sigma_n$ of *pure string motions* or *pure symmetric automorphisms* is the analogue of the pure braid group and has a natural action of the signed permutation group B_n. The presentation of the cohomology ring that we recall in Table 1 can be used to show that $H^*(P\Sigma_\bullet, \mathbb{Q})$ has the structure of what she calls a graded $FI_{BC}\#$-algebra (see [**Wil15**, Theorem 5.3]) which turns out to be finitely generated by $H^1(P\Sigma_\bullet, \mathbb{Q})$. By restricting to the S_n-action, one recovers finite generation for $H^p(P\Sigma_\bullet, \mathbb{Q})$ as FI-modules.

REMARK 4.2. The cohomology groups of P_n, PvB_n, PfB_n and $P\Sigma_n$ have the extra-structure of an $FI\#$-module. By [**CEF15**, 4.1.7(g)], the corresponding characters are given by a unique polynomial character for all $n \geq 0$. In contrast, for the spaces $M_n := \overline{\mathcal{M}}_{0,n}(\mathbb{R})$ and $\mathcal{M}_{0,n+1}$, the formulas for the character of their first cohomology do not extend for the cases $n = 0$ and $n = 1$, respectively. However, as the referee pointed out, it is interesting to notice that the formulas for $\chi_{H^1(M_n)}$ and $\chi_{H^1(\mathcal{M}_{0,n+1})}$ are accurate once the spaces considered are non-empty ($n \geq 3$ and $n \geq 2$, respectively). For instance, the formula for $\chi_{H^1(M_n)}$ is identically 0 as a class function on S_n for $n = 3$ which corresponds to the fact that $H^1(M_3; \mathbb{Q}) = 0$. And the formula for $\chi_{H^1(\mathcal{M}_{0,n+1})}$ works for all $n \geq 2$ (see for example [**JR16**, Formula (2)]).

Hyperplane complements. Another generalization of configuration spaces on the plane is given by hyperplane complements for other Weyl groups. Consider the canonical action of $\mathcal{W}_n$ on $\mathbb{C}^n$ by (signed) permutation matrices and let $\mathcal{A}(n)$ be the set of hyperplanes fixed by the (complexified) reflections of $\mathcal{W}_n$. The sequence of hyperplane complements $\mathcal{M}_{\mathcal{W}}(n) := \mathbb{C}^n - \bigcup_{H \in \mathcal{A}(n)} H$ can be encoded as a co-$FI_{\mathcal{W}}$-space $\mathcal{M}_{\mathcal{W}}(\bullet)$. In particular $\mathcal{M}_A(\bullet) = \mathcal{F}(\mathbb{C}, \bullet)$. The corresponding fundamental groups are referred as "generalized pure braid groups". The work in [**Bri73**] and [**OS80**] show that the cohomology ring for $\mathcal{M}_{\mathcal{W}}(n)$ is also generated by cohomology classes of degree one and arguments as before apply. The details are worked out in [**Wil15**, Thm 5.8].

Complex points of the Deligne-Mumford compatification of $\mathcal{M}_{0,n}$. Another related example is the complex locus of the variety $\overline{\mathcal{M}}_{0,n}$, that we denote by $\overline{\mathcal{M}}_{0,n}(\mathbb{C})$. From our discussion in Section 2, it is clear that $H^*(\overline{\mathcal{M}}_{0,\bullet}(\mathbb{C}); \mathbb{Q})$ has the struture of a graded FI-algebra. Moreover, from the results in [**Kee92**] it follows that it is generated by the FI-module $H^2(\overline{\mathcal{M}}_{0,\bullet}(\mathbb{C}); \mathbb{Q})$. However, the computations in [**Kee92**] also imply that the dimension of $H^i(\overline{\mathcal{M}}_{0,n}(\mathbb{C}); \mathbb{Q})$ is exponential in n. Therefore, from 1.2 it follows that the FI-modules $H^i(\overline{\mathcal{M}}_{0,\bullet}(\mathbb{C}); \mathbb{Q})$ are not finitely generated.

References

[Arn69] V. I. Arnol′d, *The cohomology ring of the group of dyed braids* (Russian), Mat. Zametki **5** (1969), 227–231. MR0242196

[BEER06] Laurent Bartholdi, Benjamin Enriquez, Pavel Etingof, and Eric Rains, *Groups and Lie algebras corresponding to the Yang-Baxter equations*, J. Algebra **305** (2006), no. 2, 742–764. MR2266850

[Bri73] Egbert Brieskorn, *Sur les groupes de tresses [d'après V. I. Arnol′d]* (French), Séminaire Bourbaki, 24ème année (1971/1972), Exp. No. 401, Springer, Berlin, 1973, pp. 21–44. Lecture Notes in Math., Vol. 317. MR0422674

[CEF15] Thomas Church, Jordan S. Ellenberg, and Benson Farb, *FI-modules and stability for representations of symmetric groups*, Duke Math. J. **164** (2015), no. 9, 1833–1910. MR3357185

[CEFN14] Thomas Church, Jordan S. Ellenberg, Benson Farb, and Rohit Nagpal, *FI-modules over Noetherian rings*, Geom. Topol. **18** (2014), no. 5, 2951–2984. MR3285226

[CF13] Thomas Church and Benson Farb, *Representation theory and homological stability*, Adv. Math. **245** (2013), 250–314. MR3084430

[Chu12] Thomas Church, *Homological stability for configuration spaces of manifolds*, Invent. Math. **188** (2012), no. 2, 465–504. MR2909770

[Dev99] Satyan L. Devadoss, *Tessellations of moduli spaces and the mosaic operad*, Homotopy invariant algebraic structures (Baltimore, MD, 1998), Contemp. Math., vol. 239, Amer. Math. Soc., Providence, RI, 1999, pp. 91–114. MR1718078

[DJS03] M. Davis, T. Januszkiewicz, and R. Scott, *Fundamental groups of blow-ups*, Adv. Math. **177** (2003), no. 1, 115–179. MR1985196

[DM69] P. Deligne and D. Mumford, *The irreducibility of the space of curves of given genus*, Inst. Hautes Études Sci. Publ. Math. **36** (1969), 75–109. MR0262240

[EHKR10] Pavel Etingof, André Henriques, Joel Kamnitzer, and Eric M. Rains, *The cohomology ring of the real locus of the moduli space of stable curves of genus 0 with marked points*, Ann. of Math. (2) **171** (2010), no. 2, 731–777. MR2630055

[Gai96] Giovanni Gaiffi, *The actions of S_{n+1} and S_n on the cohomology ring of a Coxeter arrangement of type A_{n-1}*, Manuscripta Math. **91** (1996), no. 1, 83–94. MR1404419

[JMM06] Craig Jensen, Jon McCammond, and John Meier, *The integral cohomology of the group of loops*, Geom. Topol. **10** (2006), 759–784. MR2240905

[JR11] Rita Jimenez Rolland, *Representation stability for the cohomology of the moduli space $\mathcal{M}_g^n$*, Algebr. Geom. Topol. **11** (2011), no. 5, 3011–3041. MR2869450

[JR15] R. Jiménez Rolland, *On the cohomology of pure mapping class groups as FI-modules*, J. Homotopy Relat. Struct. **10** (2015), no. 3, 401–424. MR3385692

[JR16] Rita Jiménez Rolland, *The cohomology of $\mathcal{M}_{0,n}$ as an FI-module*, Configuration spaces, Springer INdAM Ser., vol. 14, Springer, [Cham], 2016, pp. 313–323. MR3615738

[Kau99] Louis H. Kauffman, *Virtual knot theory*, European J. Combin. **20** (1999), no. 7, 663–690. MR1721925

[Kee92] Sean Keel, *Intersection theory of moduli space of stable n-pointed curves of genus zero*, Trans. Amer. Math. Soc. **330** (1992), no. 2, 545–574. MR1034665

[KM18] Alexander Kupers and Jeremy Miller, *Representation stability for homotopy groups of configuration spaces*, J. Reine Angew. Math. **737** (2018), 217–253, DOI 10.1515/crelle-2015-0038. MR3781336

[KV07] Joachim Kock and Israel Vainsencher, *An invitation to quantum cohomology*, Progress in Mathematics, vol. 249, Birkhäuser Boston, Inc., Boston, MA, 2007. Kontsevich's formula for rational plane curves. MR2262630

[Lee] Peter Lee, *On the action of the symmetric group on the cohomology of groups related to (virtual) braids*, Preprint (2013) arXiv:1304.4645.

[Lee13] Peter Lee, *The pure virtual braid group is quadratic*, Selecta Math. (N.S.) **19** (2013), no. 2, 461–508. MR3090235

[OS80] Peter Orlik and Louis Solomon, *Combinatorics and topology of complements of hyperplanes*, Invent. Math. **56** (1980), no. 2, 167–189. MR558866

[Rai09] Eric M. Rains, *The action of S_n on the cohomology of $\overline{M}_{0,n}(\mathbb{R})$*, Selecta Math. (N.S.) **15** (2009), no. 1, 171–188. MR2511203

[Wil14] Jennifer C. H. Wilson, $\mathrm{FI}_{\mathcal{W}}$*-modules and stability criteria for representations of classical Weyl groups*, J. Algebra **420** (2014), 269–332. MR3261463

[Wil15] Jennifer C. H. Wilson, $\mathrm{FI}_{\mathcal{W}}$*-modules and constraints on classical Weyl group characters*, Math. Z. **281** (2015), no. 1-2, 1–42. MR3384861

Instituto de Matemáticas, Universidad Nacional Autónoma de México, Oaxaca de Juárez, Oaxaca, México 68000

Email address: rita@im.unam.mx

CINVESTAV, Colonia Pedro Zacatenco, Ciudad de México, México 07360

Email address: jmaya@math.cinvestav.mx

Contemporary Mathematics
Volume **709**, 2018
http://dx.doi.org/10.1090/conm/709/14292

Two proofs of Taubes' theorem on strictly ergodic flows

Victor Kleptsyn and Ana Rechtman

ABSTRACT. The purpose of this note is to give two new proofs of C. H. Taubes' result concerning the helicity of a strictly ergodic flow. A strictly ergodic flow is a uniquely ergodic volume preserving flow. The theorem states that on a 3-manifold the helicity of such flows is zero. For both proofs we use C. H. Taubes' result on the existence of periodic orbits for Reeb vector fields [J. Gökova Geom. Topol. 13 (2009), pp. 1337–1417] and for the second one we also use the characterization of geodesible vector fields by D. Sullivan [J. Pure Appl. Algebra 13 (1978), pp. 101–104].

1. Introduction

Let X be a vector field on a closed oriented 3-manifold M, that preserves the volume form Ω. Thus the Lie derivative $L_X\Omega = 0$. Cartan's formula implies that $\omega = \iota_X\Omega$ is a closed 2-form. The vector field X is exact if ω is exact. In this case, let α be a primitive of ω. Observe that in $\mathbb{S}^3$ or in any homotopy sphere, any volume preserving vector field is exact. The helicity of the vector field X is defined as the integral

$$h(X) = \int \alpha \wedge d\alpha,$$

that does not depends on the choice of the 1-form α. The helicity is an important invariant in the study of fluid flows. When the manifold is simply connected the helicity coincides with the asymptotic linking number, as proved by Arnold [**1**].

A vector field is uniquely ergodic if its flow admits a unique invariant measure. When such a measure is a smooth volume, equivalently a volume form, the vector field is strictly ergodic.

THEOREM 1.1 (Taubes, [**6**]). *Let X be a strictly ergodic exact vector field on a closed 3-manifold M. Then $h(X) = 0$.*

Let f be the function defined by $f = \alpha(X)$ or equivalently $\alpha \wedge d\alpha = f\Omega$. Observe that $h(X)$ is the integral of f with respect to Ω. The proofs we present here give the following stronger statement:

THEOREM 1.2. *If X is a minimal exact vector field preserving the volume form Ω, then there exists an invariant measure μ such that $\int_M f\, d\mu = 0$.*

2010 *Mathematics Subject Classification.* Primary 37C40; Secondary 57M27.
Key words and phrases. Helicity, strictly ergodic flows.

By minimal we understand a vector field whose orbits are dense in M. A strictly ergodic vector field is minimal. The minimality or strict ergodicity is used in different ways in the proofs below. In the first proof, it is used to say that X is not a Reeb vector field and thus find two invariant measures for which the integral of f is strictly positive with respect to the first measure and strictly negative with respect to the second measure. A linear combination of these invariant measures gives the measure μ in Theorem 1.2.

In the second poof, the minimality or strict ergodicity of X is used to say that X is not geodesible (in other words, X is not the Reeb vector field of a stable Hamiltonian structure). D. Sullivan's characterisation of geodesible vector fields gives the existence of the measure μ. We thank the referee for pointing out a proof of Theorem 1.1 that was explained to him by Gabriel Paternain in 2008 and differs little from the second proof in this paper. In fact, instead of using the characterisation by D. Sullivan of geodesible vector fields, one can use Theorem 5.2 of D. McDuff's article [**3**].

C. H. Taubes' proof of Theorem 1.1 relies on Seiberg-Witten equations. He uses this technique in a similar way as for proving the existence of a periodic orbit of a Reeb vector field on a 3-manifold [**7**]. The proofs presented here use the fact that a Reeb vector field has a periodic orbit and hence cannot be minimal. In the case of $\mathbb{S}^3$ or a 3-manifold with non-trivial second homotopy group, H. Hofer proved the existence of periodic orbit for Reeb vector fields using pseudoholomorphic curves [**2**].

There are two types of examples of strictly ergodic flows on 3-manifolds. First, suspensions of strictly ergodic diffeomorphisms of the 2-torus. Second quotients of the horocycle flow. The following example was explained to us by A. Verjovsky. Consider the Brieskorn manifolds

$$V_{p,q,r} = \{(z_1, z_2, z_3) \in \mathbb{C}^3 \mid z_1^p + z_2^q + z_3^r = 0,\ z_1^2 + z_2^2 + z_3^2 = 1\},$$

with p, q, r positive integers and two by two relative primes. Then $V_{p.q.r}$ is a homology sphere and $\mathbb{S}^3$ if p, q or r is equal to 1. If $\frac{1}{p} + \frac{1}{q} + \frac{1}{r} < 1$ then $V_{p,q,r}$ is a quotient of $SL(2,\mathbb{R})$ by a lattice. The horocycle flow on $SL(2,\mathbb{R})$ defines a flow in the quotient space that is a strictly ergodic flow on the 3-manifold. The horocycle flow defines a strictly ergodic flow on any compact quotient of $SL(2,\mathbb{R})$, but these quotients are not always homology spheres.

2. First proof

Consider an exact vector field X preserving the volume form Ω, then, using the same notation as in the introduction, write $\iota_X \Omega = \omega = d\alpha$ for a differential 1-form α. In this section we prove the following result:

THEOREM 2.1. *Let X be an exact volume preserving vector field. Assume that X has no periodic orbits and let f be the function defined by $\alpha \wedge d\alpha = f\Omega$. Then, there exists an invariant measure μ with $\int f\, d\mu = 0$.*

Clearly, Theorems 1.2 and 1.1 can be deduced from Theorem 2.1.

Before starting the proof, recall that a differential 1-form λ on a closed oriented 3-manifold M is a contact form if $\lambda \wedge d\lambda \neq 0$. For such a form, its Reeb vector field X is defined by the equations $\lambda(X) = 1$ and $\iota_X d\lambda = 0$. Thus X preserves λ and the volume form $\lambda \wedge d\lambda$.

PROOF. Let ϕ_t be the flow of the vector field X. The conclusion in Theorem 2.1 can be reformulated in the following way: it cannot happen that the integrals of f with respect to all the ϕ_t-invariant measures are all strictly positive or all strictly negative.

Assume the converse and let all such integrals be, for instance, strictly positive. Consider the family of 1-forms

$$\alpha_t = \frac{1}{t}\int_0^t \phi_\tau^* \alpha \, d\tau,$$

obtained by averaging α along the flow. Since the flow preserves Ω, it preserves $\omega = \iota_X \Omega$ and hence for any t the 1-form α_t is a primitive of ω with

$$\alpha_t \wedge \omega = \left(\frac{1}{t}\int_0^t f \circ \phi_\tau \, d\tau\right)\Omega.$$

Notice that among the time average functions $\bar{f}_t = \frac{1}{t}\int_0^t f \circ \phi_\tau \, d\tau$, there is a strictly positive function. Indeed, if this was not the case, there would exists a sequence of points p_{t_n} such that $\bar{f}_{t_n}(p_{t_n}) \leq 0$. But,

$$\bar{f}_{t_n}(p_{t_n}) = \frac{1}{t_n}\int_0^{t_n} f(\phi_\tau(p_{t_n}))\, d\tau = \int f \, d\mu_n,$$

where

$$\mu_n = \frac{1}{t_n}\int_0^{t_n} \delta_{\phi_\tau(p_{t_n})}\, d\tau$$

the time average of the Dirac measure $\delta_{p_{t_n}}$. Consider a Riemannian metric on M such that the norm of X is identically 1, then the length of the orbit segment from p_{t_n} to $\phi_{t_n}(p_{t_n})$ is equal to t_n. Thinking the measures μ_n as 1-currents, the mass of μ_n is 1. Extracting a weakly converging subsequence μ_{n_k} of more and more invariant measures, we find that the limit μ is a non-zero invariant measure with

$$\int f \, d\mu = \lim_{k\to\infty} \int f \, d\mu_{n_k} \leq 0,$$

giving us a contradiction to our assumption. Hence for some t we have $\alpha_t \wedge d\alpha_t = \bar{f}_t \Omega$, with

$$\bar{f}_t = \frac{1}{t}\int_0^t f \circ \phi_\tau \, d\tau > 0.$$

But then α_t is a contact form and $\alpha_t(X) = \bar{f}_t > 0$ and $\iota_X d\alpha_t = \iota_X \omega = 0$. Thus, modulo a reparametrisation by a positive function, X is the Reeb vector field of α_t. Due to Taubes' result, X possesses a periodic orbit. This contradiction concludes the proof. □

3. Second proof

A non-singular vector field is geodesible if there exists a Riemannian metric on the ambient manifold making all the orbits geodesics. As remarked in [**5**], a vector field is geodesible if and only if there exists a differential 1-form λ such that $\lambda(X) = 1$ and $\iota_X d\lambda = 0$. Observe that geodesible vector fields form a larger family than Reeb vector fields, since the form λ might be a closed 1-form. Geodesible volume preserving vector fields are also known in the literature as Reeb vector fields of stable Hamiltonian structures.

THEOREM 3.1 (Sullivan [**5**]). *A vector field is geodesible if and only if any foliation cycle cannot be approximated by tangent 2-chains.*

A foliation cycle is a flat current that is obtained from positive linear combinations of the Dirac currents defined by X. We recall that foliation cycles are in one to one correspondence with the invariant measures of the flow of X, as proved by D. Sullivan [**4**]. The term tangent 2-chain means a flat current of integration on sets that are tangent to the vector field X, this implies that the 2-chain evaluated on any 2-form whose kernel contains X must be equal to zero. We can now give the second proof to Theorem 1.1.

PROOF. We use the previous notation, that is Ω is the invariant volume form and $\iota_X\Omega = \omega = d\alpha$.

If the flow of X is strictly ergodic then it is a minimal flow: every orbit is dense. Assume that X is geodesible, then there exists a differential 1-form λ such that $\lambda(X) = 1$ and $\iota_X d\lambda = 0$. In particular, λ and $\lambda \wedge d\lambda$ are invariant forms. Since the latter is a differential 3-form we can write $\lambda \wedge d\lambda = g\Omega$ for some function g. Since both Ω and $\lambda \wedge d\lambda$ are invariant forms, the function g is invariant and by the minimality of the flow it is constant. If $g \neq 0$, the vector field X is the Reeb vector field of the contact form λ, which contradicts minimality by the existence of a periodic orbit [**7**]. If $g = 0$, observe that $d\lambda = 0$ since X is not in the kernel of λ. Then λ is a closed 1-form positive on X. Tischler's theorem [**8**] implies that M fibers over the circle and each fiber is a global section for the flow. This means, that the fibers are closed 2-manifold S transverse to the vector field and the flow is the suspension of a diffeomorphism of S. This is impossible, an exact vector field cannot admit a global section as a consequence of Stokes' theorem:

$$0 = \int_{\partial S} \alpha = \int_S d\alpha = \int_S \omega > 0,$$

where the last inequality follows since ω is an area form on S.

Thus X is not geodesible. Since X is strictly ergodic and invariant measures are in one to one correspondence with foliation cycles [**4**], there is only one foliation cycle Z that corresponds to the volume form Ω. Then

$$Z(\alpha) = \int_M \alpha \wedge d\alpha = \int_M f\Omega = h(X),$$

as remarked in page 232 of [**4**]. Then by Theorem 3.1 there exists a sequence of tangent 2-chains B_n such that ∂B_n tends to Z. Since X is in the kernel of $d\alpha$ and is tangent to B_n, we have that $B_n(d\alpha) = 0$. Thus,

$$Z(\alpha) = \lim_{n\to\infty} \partial B_n(\alpha) = \lim_{n\to\infty} B_n(d\alpha) = 0.$$

□

REMARK 3.2. If X is minimal but not strictly ergodic, the proof above implies that X is not geoedesible. Then there is a the foliation cycle approximated by the boundaries of tangent 2-chains, that is not necessarily the foliation cycle associated to the invariant volume form. Thus the proof above provides the existence of an invariant measure μ for which $\int_M f\, d\mu = 0$, proving Theorem 1.2.

References

[1] Vladimir I. Arnold and Boris A. Khesin, *Topological methods in hydrodynamics*, Applied Mathematical Sciences, vol. 125, Springer-Verlag, New York, 1998. MR1612569

[2] Helmut Hofer, *Pseudoholomorphic curves in symplectizations with applications to the Weinstein conjecture in dimension three*, Invent. Math. **114** (1993), no. 3, 515–563. MR1244912

[3] Dusa McDuff, *Applications of convex integration to symplectic and contact geometry* (English, with French summary), Ann. Inst. Fourier (Grenoble) **37** (1987), no. 1, 107–133. MR894563

[4] Dennis Sullivan, *Cycles for the dynamical study of foliated manifolds and complex manifolds*, Invent. Math. **36** (1976), 225–255. MR0433464

[5] Dennis Sullivan, *A foliation of geodesics is characterized by having no "tangent homologies"*, J. Pure Appl. Algebra **13** (1978), no. 1, 101–104. MR508734

[6] Clifford Henry Taubes, *An observation concerning uniquely ergodic vector fields on 3-manifolds*, J. Gökova Geom. Topol. GGT **3** (2009), 9–21. MR2595753

[7] Clifford Henry Taubes, *The Seiberg-Witten equations and the Weinstein conjecture. II. More closed integral curves of the Reeb vector field*, Geom. Topol. **13** (2009), no. 3, 1337–1417. MR2496048

[8] David Tischler, *On fibering certain foliated manifolds over* S^1, Topology **9** (1970), 153–154. MR0256413

CNRS, Institut de recherche mathématique de Rennes, Université de Rennes 1, Campus de Beaulieu, 35042 Rennes cedex, France
Email address: kleptsyn@gmail.com

Institut de Recherche Mathématique Avancée, Université de Strasbourg, 7 rue René Descartes, 67084 Strasbourg cedex, France
Email address: rechtman@math.unistra.fr

Contemporary Mathematics
Volume **709**, 2018
http://dx.doi.org/10.1090/conm/709/14293

The fibers of the Prym map of étale cyclic coverings of degree 7

Herbert Lange and Angela Ortega

ABSTRACT. We study the Prym varieties arising from étale cyclic coverings of degree 7 over a curve of genus 2. We prove that these Prym varieties are products of Jacobians $JY \times JY$ of genus 3 curves Y with polarization type $D = (1,1,1,1,1,7)$. We describe the fibers of the Prym map between the moduli space of such coverings and the moduli space of abelian sixfolds with polarization type D, admitting an automorphism of order 7.

1. Introduction

Given a finite covering $f : \widetilde{C} \to C$ between smooth projective curves, one can associate to f an abelian subvariety of the Jacobian $J\widetilde{C}$ by taking the connected component containing the zero of the kernel of the norm map $\mathrm{Nm}_f : J\widetilde{C} \to JC$, $[D] \mapsto [f_*(D)]$. The resulting variety

$$P(f) := (\mathrm{Ker}\, \mathrm{Nm}_f)^0 \subset J\widetilde{C}$$

is called the *Prym variety of* f. The restriction of the theta divisor of $J\widetilde{C}$ to $P(f)$ defines a polarization on the Prym variety which is known to be twice a principal polarization when f is an étale double covering (see [**9**]). The assignment $[f : \widetilde{C} \to C] \mapsto P(f)$ yields a *Prym map* between the moduli space of the corresponding coverings and the moduli space of abelian varieties of suitable dimension and polarization (not necessarily principally polarized). In very few cases the Prym maps are generically finite and dominant over its image (see [**1**], [**5**], [**6**], [**7**] and [**8**]) and often the structure of the fibers can be understood in geometrical terms ([**4**], [**6**]).

We study the Prym varieties associated to a non-trivial 7 cyclic étale covering $f : \widetilde{C} \to C$ over a curve C of genus 2. The corresponding Prym variety $P(f)$ is a 6-dimensional abelian variety with polarization type $D = (1,1,1,1,1,7)$. Let $\mathcal{R}_{2,7}$ denote the moduli space of these coverings and $\mathcal{B}_D$ the moduli space of abelian varieties of dimension 6 with a polarization of type D and an automorphism of order 7 compatible with the polarization. In [**8**] we proved that the Prym map $\mathrm{Pr}_{2,7} : \mathcal{R}_{2,7} \to \mathcal{B}_D$ is dominant and since both moduli spaces are 3-dimensional (see

2010 *Mathematics Subject Classification.* Primary 14H40, 14H30.
Key words and phrases. Prym variety, Prym map.
The second author was partially supported by the *Sonderforschungsbereich* 647 "Raum-Zeit-Materie".

[**8**] for a proof), the map $\mathrm{Pr}_{2,7}$ is generically finite and in fact, of degree 10. In this article we give a geometric description of the fibers of the Prym map $\mathrm{Pr}_{2,7}$.

First we prove in Section 2 that for all $[f : \widetilde{C} \to C] \in \mathcal{R}_{2,7}$ the associated Prym variety $P(f) = JY \times JY$, where Y is a curve of genus 3 whose Jacobian admits a multiplication in the totally real cubic subfield $\mathbb{Q}(\zeta_7)^0$ of the cyclotomic field $\mathbb{Q}(\zeta_7)$ (Proposition 2.4); moreover Y is uniquely determined by the Prym variety. In Section 3 we show that the curve Y admits a degree 7-covering $Y \to \mathbb{P}^1$ with ramification type $(2,2,2,1)$, that is, f is simply ramified over each branch point at 3 points and unramified at 1. Let $\mathcal{C}_3$ denote the locus in $\mathcal{M}_3$ of curves Y having a covering of this type. The main result is the following.

THEOREM 1.1. *The elements of a generic fiber of the Prym map* $\mathrm{Pr}_{2,7} : \mathcal{R}_{2,7} \to \mathcal{B}_D$ *are in bijection with the set of degree 7 coverings* $\overline{f} : Y \to \mathbb{P}^1$ *with ramification type* $(2,2,2,1)$ *over 6 general points that a curve* $Y \in \mathcal{C}_3$ *admits.*

REMARK 1.2. In [**8**] we considered a partial compactification of $\mathrm{Pr}_{2,7}$ which is proper and generically finite of degree 10. So, a finite fiber consists of 10 elements counted with multiplicities. This fact together with Theorem 1.1 shows that the number of degree 7 coverings $\overline{f} : Y \to \mathbb{P}^1$ with ramification type $(2,2,2,1)$ over each of the 6 ramification points is ≤ 10.

Acknowledgments. We would like to thank Alice Silverberg and Karl Rubin for their help with the proof of Lemma 2.7.

2. Description of the Prym variety as product of Jacobians

Let C be a genus 2 curve and $f : \widetilde{C} \to C$ a non-trivial étale cyclic covering of degree 7. We shall give an explicit description of the associated Prym variety $P(f)$. Part of the results in this section are contained in [**10**]. It is known that the hyperelliptic involution ι of C lifts, though non-canonically, to an involution $j : \widetilde{C} \to \widetilde{C}$. If we denote by σ the automorphism of order 7 on the cover $\widetilde{C}$, then j, σ generate the dihedral group

$$D_7 = \langle j, \sigma \mid j^2 = \sigma^7 = 1, j\sigma j = \sigma^{-1} \rangle.$$

Moreover, the composed map $\widetilde{C} \to \mathbb{P}^1$ is Galois with group D_7 ([**3**, Proposition 2.1]). Using the commutativity of the diagram

$$\begin{array}{ccc} \widetilde{C} & \xrightarrow{\;j\;} & \widetilde{C} \\ {\scriptstyle f}\big\downarrow & & \big\downarrow{\scriptstyle f} \\ C & \xrightarrow{\;\iota\;} & C \end{array} \tag{2.1}$$

one checks that the images of the fixed points of j are precisely the fixed points of ι and, since f is étale, there is only one fixed point on each fiber of a fixed point of ι. Let

$$g : \widetilde{C} \to Y := \widetilde{C}/\langle j \rangle$$

be the double covering associated to the involution j. Then g is ramified exactly at the 6 fixed points of j, so by Hurwitz formula Y is of genus 3. Notice that since g is a ramified cover, the map $g^* : JY \to J\widetilde{C}$ is injective, so we can regard JY as subvariety in $J\widetilde{C}$. As before, let $P = P(f)$ be the Prym variety associated to f.

PROPOSITION 2.1. *The map*

$$\begin{array}{rcl} \psi : JY \times JY & \to & P \\ (x,y) & \mapsto & x + \sigma(y) \end{array}$$

is an isomorphism of abelian varieties.

Certainly ψ does not respect the canonical polarizations, since $JY \times JY$ is canonically principally polarized whereas the induced polarization Ξ on P is of type $(1, \dots, 1, 7)$.

PROOF. First we check that JY and $\sigma(JY)$ are contained in P. Recall that $P = \mathrm{Ker}(1 + \sigma + \cdots + \sigma^6)^0$ and $g^*(JY) = \mathrm{Im}(1 + j)$. On $J\widetilde{C}$

$$(1 + \cdots + \sigma^6)(1 + j) = 1 + \cdots + \sigma^6 + j + \cdots + j\sigma^6 = 0$$

since this map is the the norm map of $\widetilde{C} \to \mathbb{P}^1$. This shows that $JY \subset P$ and as σ acts on P, we also have $\sigma(JY) \subset P$.

Note that $\dim(JY \times JY) = \dim P$, so in order to prove the proposition it suffices to show that ψ is injective. Recall that, since $Y = \widetilde{C}/\langle j \rangle$, the elements of $JY \simeq g^*JY$ are fixed by j. Let $(x, y) \in JY \times JY$ such that $x + \sigma(y) = 0$. Then

$$x = j(x) = -j\sigma(y) = -\sigma^6 j(y) = -\sigma^6(y)$$

and hence $\sigma^2(x) = -\sigma(y) = x$. This shows that $x \in \mathrm{Fix}(j, \sigma)$. In particular $x \in \mathrm{Fix}(\sigma) \cap P \subset J\widetilde{C}[7]$. Since $\sigma(x) = x$ and f is étale there exists an element $z \in JC$ such that $x = f^*(z)$. Further, $j(x) = x$ and the commutativity of diagram (2.1) implies that

$$x = j(x) = jf^*(z) = f^*\iota(z) = -f^*(z) = -x$$

since on JC we have $\iota(z) = -z$, that is $x \in J\widetilde{C}[2]$. In conclusion $x \in J\widetilde{C}[2] \cap J\widetilde{C}[7] = \{0\}$ and therefore ψ is injective. $\square$

In order to describe the polarization on the product, let Ξ denote the polarization of type $(1,1,1,1,1,7)$ on P. Hence its pull back $\psi^*\Xi$ is also of type $(1,1,1,1,1,7)$. Now Ξ is the restriction of the canonical polarization of $J\widetilde{C}$ to P. So we may consider the polarization $\psi^*\Xi$ as the pullback of the canonical polarization of $J\widetilde{C}$ to $JY \times JY$.

We identify $\widehat{J\widetilde{C}} = J\widetilde{C}$ and $\widehat{JY} = JY$ via the canonical polarizations. The induced split polarization on $JY \times JY$ gives an identification $(JY \times JY)^\wedge = JY \times JY$. Let $\theta : JY \times JY \xrightarrow{\psi} P \hookrightarrow J\widetilde{C}$ be the embedding. For $i = 1, 2$ denote by θ_i the composition

$$\theta_i : JY \to JY \times JY \xrightarrow{\theta} J\widetilde{C},$$

where the first map is the natural embedding into the i-th factor. One checks that

$$\theta_1 = g^* : JY \to J\widetilde{C} \quad \text{and} \quad \theta_2 = \sigma \circ g^* : JY \to J\widetilde{C}.$$

Since the dual of g^* is the norm map

$$N_g = 1 + j : J\widetilde{C} \to JY,$$

this implies that

$$\widehat{\theta_1} = \widehat{g^*} = N_g = 1 + j : J\widetilde{C} \to JY \quad \text{and} \quad \widehat{\theta_2} = \widehat{g^*} \circ \sigma^{-1} = \sigma^6 + \sigma j : J\widetilde{C} \to JY,$$

since $\widehat{\sigma} = \sigma^{-1}$.

Now note that $\widehat{\theta}_1\theta_1 = N_g g^* = 2_{JY}$ and $\widehat{\theta}_2\theta_2 = N_g\sigma^{-1}\sigma g^* = 2_{JY}$. Hence the matrix of $\phi_{\psi^*\Xi} : JY \times JY \to JY \times JY$ is

$$\phi_{\psi^*\Xi} = \begin{pmatrix} \widehat{\theta}_1\theta_1 & \widehat{\theta}_1\theta_2 \\ \widehat{\theta}_2\theta_1 & \widehat{\theta}_2\theta_2 \end{pmatrix} = \begin{pmatrix} 2_{JY} & N_g\sigma g^* \\ N_g\sigma^{-1}g^* & 2_{JY} \end{pmatrix}.$$

So we have,

PROPOSITION 2.2. *Let $g : \widetilde{C} \to Y = \widetilde{C}/\langle j\rangle$ be the natural map and suppose $\widehat{J\widetilde{C}} = J\widetilde{C}$ and $\widehat{JY} = JY$ via the canonical principal polarizations. If we denote by φ the isogeny $\varphi = N_g\sigma g^* : JY \to JY$, then the polarization $\psi^*\Xi$ on $JY \times JY$ is given by the matrix*

$$\phi_{\psi^*\Xi} = \begin{pmatrix} 2_{JY} & \varphi \\ \widehat{\varphi} & 2_{JY} \end{pmatrix}.$$

In order to study the polarizations of type $(1,\ldots,1,7)$ on the product $P := JY \times JY$, we recall from [**2**, Section 5.2] the description of the set of polarizations of degree 7 on P. According to [**2**, Theorem 5.2.4] the canonical principal polarization on $JY \times JY$ induces a bijection between the sets of

(a) polarizations of degree 7 on $JY \times JY$ and

(b) totally positive symmetric endomorphism of $JY \times JY$ with analytic norm 7.

The endomorphisms of $JY \times JY$ are given by square matrices of degree 2 with entries in $\mathrm{End}(JY)$. The symmetry in (b) is with respect to the Rosati involution. Since the Rosati involution with respect to the canonical polarization of $JY \times JY$ is just the transposition of matrices, we are looking for the symmetric positive definite matrices

$$A := \begin{pmatrix} \rho_a(\alpha) & \rho_a(\beta) \\ \rho_a(\beta)^t & \rho_a(\delta) \end{pmatrix}$$

with determinant 7, where $\alpha, \beta, \delta \in \mathrm{End}(JY)$.

PROPOSITION 2.3. *If P admits a polarization of degree 7, then* $\mathrm{End}(JY) \supsetneqq \mathbb{Z}$.

PROOF. Suppose $\mathrm{End}(JY) = \mathbb{Z}$. For a general Y the Rosati involution on $JY \times JY$ is the transposition composed with the Rosati involution of the pieces JY, but for a general Y the Rosati involution on JY is the identity. Let A be an endomorphism of degree 7, given by the matrix A. Then α, β and δ are integers and $\rho_a(\alpha) = \alpha_{\mathbb{C}^3} = diag(\alpha,\alpha,\alpha)$ etc. and we have

$$7 = \det A = \det \begin{pmatrix} diag(\alpha,\alpha,\alpha) & diag(\beta,\beta,\beta) \\ diag(\beta,\beta,\beta) & diag(\delta,\delta,\delta) \end{pmatrix} = (\alpha\delta - \beta^2)^3.$$

Since this equation does not admit an integer solution, this gives a contradiction. □

Recall that $\mathcal{R}_{2,7}$ is an irreducible 3-dimensional variety and the Prym map is generically finite onto $\mathcal{B}_D$. So also $\mathcal{B}_D$ is of dimension 3. Since every element of $\mathcal{B}_D$ is isomorphic to $JY \times JY$ for some curve Y of genus 3, and since the number of decompositions $P = JY \times JY$ is at most countable, we get a 3-dimensional algebraic set, say $\mathcal{V}$, of curves Y such that $JY \times JY$ admits a polarization of degree 7. In fact, as a consequence of the results in the paper $\mathcal{V}$ is even a variety.

PROPOSITION 2.4. *The Jacobians JY of all curves $Y \in \mathcal{V}$ admit real multiplication in the totally real cubic number subfield $\mathbb{Q}(\zeta_7)^0$ of the cyclotomic field $\mathbb{Q}(\zeta_7)$.*

PROOF. Since by Proposition 2.3 $\mathrm{End}_{\mathbb{Q}}(JY) \neq \mathbb{Q}$, the Jacobian admits either real, quaternion or complex multiplication (in the more general sense of [**2**, Section 9.6]). But an abelian variety with quaternion multiplication is even dimensional ([**2**, 9.4 and 9.5]). If JY admits complex multiplication by a skew field of degree d^2 over a totally complex quadratic extension of a totally real number field of degree e_0, we would have (see [**2**, 9.6]) $3 = d^2 e_0 m$ for some integer $m \geq 1$. So $d = 1$ and $e_0 = 3$ (by Proposition 2.3). Then JY admits complex multiplication by a number field. Since there are only countably many such abelian varieties, we conclude that JY admits multiplication by a totally real number field F of degree say e.

Then we have $3 = em$ for some positive integer m [**2**, §9.2] and Proposition 2.3 implies $e = 3$. So $JY \times JY$ admits multiplication in $\mathrm{SL}_2(F)$, with F a totally real cubic number field. On the other hand P admits a multiplication in the cyclotomic field $\mathbb{Q}(\zeta_7)$ and this is a subfield of $\mathrm{SL}_2(F)$ only if F is the totally real cubic subfield $\mathbb{Q}(\zeta_7)^0$ in $\mathbb{Q}(\zeta_7)$. Therefore $F = \mathbb{Q}(\zeta_7)^0$. □

As a consequence of the description of the polarization in Proposition 2.1 we have

PROPOSITION 2.5. *Let $\mathcal{O}$ denote the maximal order of the totally real cubic field* $\mathrm{End}_{\mathbb{Q}}(JY)$ *and $\varphi : JY \to JY$ the isogeny of Proposition* 2.2. *The following equation admits a solution in $\mathcal{O}$*

$$7 = \det(4 \cdot \mathbf{1} - \rho_{\mathbf{a}}(\varphi\widehat{\varphi})).$$

THEOREM 2.6. *For a general covering f the decomposition $P(f) \simeq JY \times JY$ is unique up to automorphisms.*

PROOF. We may assume that the the ring of endomorphisms of JY is the maximal order $\mathcal{O}$ of the field F, since both families, the family of Prym varieties $P(f)$ and the family of Jacobians with multiplication in F are irreducible of the same dimension 3. So

$$\mathrm{End}(P(f)) \simeq M_2(\mathcal{O}),$$

the ring of matrices of degree 2 with entries in $\mathcal{O}$. Let A be any direct factor of $P(f)$. So there is an abelian subvariety B of $P(f)$ (necessarily isogenous to A) with

$$P(f) \simeq A \times B.$$

We have to show that there is an automorphism α of $P(f)$ with $\alpha(JY) = A$.

Recall that an element $\epsilon \in \mathrm{End}(P(f))$ is idempotent if $\epsilon^2 = \epsilon$. Now the direct factors of $P(f)$ correspond bijectively to the non-trivial idempotents of the endomorphism ring of $P(f)$, i.e. of $M_2(\mathcal{O})$. Namely, if A is a direct factor, then the composition ϵ of the maps

$$P(f) \simeq A \times B \xrightarrow{p_1} A \xrightarrow{i_1} A \times B \simeq P(f)$$

is an idempotent of $\mathrm{End}(P(f))$. Here p_1 and i_1 are the natural projection and inclusion. Conversely, if ϵ is an idempotent, the factor A is given by the kernel of the endomorphism $1 - \epsilon$. Moreover, if $\epsilon' = \alpha\epsilon\alpha^{-1}$ with an automorphism α of $P(f)$, then

$$\alpha(1 - \epsilon)\alpha^{-1} = 1 - \epsilon'.$$

Hence ϵ and ϵ' correspond to isomorphic direct factors. It suffices to show that all nontrivial idempotents (i.e. different from 0 and 1) are conjugate to each other, which is the content of the following lemma. □

LEMMA 2.7. *Any 2 nontrivial idempotents of $M_2(\mathcal{O})$ are conjugate to each other.*

PROOF. Let $\epsilon \in M_2(\mathcal{O})$ be an idempotent. Its minimal polynomial $p(x)$ is different from x and $x-1$, since ϵ is nontrivial.

Let $M := \mathcal{O}^2$ with ϵ acting by multiplication in the natural way. Let N_0 be its 0-eigenspace, i.e. the kernel of ϵ and N_1 its 1-eigenspace, i.e. the kernel of $\epsilon - 1$. We have

$$\epsilon M \subset N_1 \quad \text{and} \quad (\epsilon - 1)M \subset N_0,$$

since $\epsilon^2 - \epsilon$ annihilates M. Moreover,

$$N_0 \cap N_1 = 0 \tag{2.2}$$

$$N_0 + N_1 = M. \tag{2.3}$$

The first equation is clear. For the second equation note that every $m \in M$ can be expressed as $m = \epsilon m - (\epsilon - 1)m$.

Neither N_0 nor N_1 can equal M, since otherwise we would have $p(x) = x$ or $x - 1$. Hence by (2.3) neither can be zero, so N_0 and N_1 are both $\mathcal{O}$-modules of rank 1. Since $\mathcal{O}$ is a principal ideal domain, there are $m_0, m_1 \subset M$ such that

$$N_0 = \mathcal{O}m_0 \quad \text{and} \quad N_1 = \mathcal{O}m_1.$$

It follows from (2.2) and (2.3) that $\{m_0, m_1\}$ is an $\mathcal{O}$-basis of M. With respect to this basis ϵ has the form

$$\begin{pmatrix} 0 & 0 \\ 0 & 1 \end{pmatrix}.$$

This implies the assertion. □

REMARK 2.8. The proof works more generally for $M_2(\mathcal{O})$ with $\mathcal{O}$ any principal ideal domain.

3. The number of isomorphism classes of coverings $\overline{f}$

Let $f : \widetilde{C} \to C$ be an étale cyclic covering of degree 7 of a smooth curve of genus 2. As we saw in Section 2, the lifting j of the hyperelliptic involution ι of C is not unique and in fact there are exactly 7 liftings $j_0, \dots, j_6$ which are the 7 involutions of D_7. If $g_i : \widetilde{C} \to Y_i := \widetilde{C}/\langle j_i \rangle$ denotes the quotient map given by the involution $j_i := j\sigma^i$ (so $g_0 = g$) we have the following cartesian diagram.

(3.1)

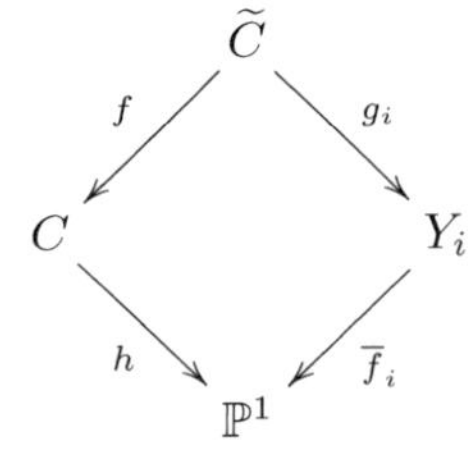

In the previous section we saw that Y_i is of genus 3 and g_i is ramified exactly at one point over each Weierstrass point of C. Since f is étale, the commutativity of (3.1) implies that the branch points of $\overline{f}_i : Y_i \to \mathbb{P}^1$ coincide with the branch points of h and each branch point is of ramification type $(2,2,2,1)$, i.e $\overline{f}_i$ is simply ramified at 3 points and unramified at 1 point over the branch point. So the branch points of $\overline{f}_i$ coincide for all i. If $p_1, \dots, p_6 \in \mathbb{P}^1$ are the branch points, then the permutations corresponding to the fibers $(\overline{f}_i)^{-1}(p_j)$ are pairwise conjugate within D_7 for all $i = 0, \dots, 6$ and fixed j with the same conjugation for $j = 1, \dots, 6$ and such that the non-ramified points of f_j over each p_i are pairwise different for $j = 0, \dots, 6$.

Note that the coverings $\overline{f}_i$ correspond to conjugate subgroups of D_7. Hence the $\overline{f}_i : Y_i \to \mathbb{P}^1$ are isomorphic coverings.

LEMMA 3.1. *For any general smooth curve C of genus 2 with double covering $h : C \to \mathbb{P}^1$, there is a canonical bijection between the sets of*

(a) *isomorphism classes of étale cyclic coverings $f : \widetilde{C} \to C$ of degree 7,*
(b) *isomorphism classes of degree-7 coverings $\overline{f} : Y \to \mathbb{P}^1$ of ramification type $(2,2,2,1)$ over each of the 6 ramification points.*

PROOF. We saw above that any covering f in (a) gives 7 coverings in (b), which are all isomorphic to each other. Conversely, let $\overline{f} : Y \to \mathbb{P}^1$ be one of the coverings in (b). Since C is general, the monodromy group G of $\overline{f}$ coincides with the Galois group of the Galois closure of $Y \to \mathbb{P}^1$. Define

$$\widetilde{C} := C \times_{\mathbb{P}^1} Y.$$

According to the Lemma of Abhyankar [**11**, Lemma 2.14] the projection $\widetilde{C} \to C$ is étale. So $\widetilde{C}$ is smooth and it is easy to see that it is Galois over $\mathbb{P}^1$ with Galois group $\simeq D_7$. Hence $\widetilde{C}$ is the Galois closure of $\overline{f}$. In particular $\widetilde{C} \to C$ is cyclic étale of degree 7. Clearly both constructions are inverse to each other and isomorphic coverings in (a) correspond to isomorphic coverings in (b). □

Given a covering $f : \widetilde{C} \to C$ corresponding to a general element in $\mathcal{R}_{2,7}$, by Lemma 3.1 there is a corresponding degree 7 map $Y \to \mathbb{P}^1$ of ramification type $(2,2,2,1)$ and, according to Proposition 2.1, the Prym variety $P(f)$ decomposes as

$$P(f) = JY \times JY.$$

Now there are $7^4 - 1 = 2400$ non-trivial 7-torsion points of C. Since a cyclic étale cover $\widetilde{C} \to C$ of degree 7 is given by a cyclic subgroup of order 7 of $JC[7]$, there are exactly

$$\frac{2400}{6} = 400$$

isomorphism classes of cyclic étale coverings $\widetilde{C} \to C$.

Let $p_1, \dots, p_6$ be the branch points of the hyperelliptic covering $C \xrightarrow{h} \mathbb{P}^1$. Let N be the number of isomorphism classes of coverings $\overline{f} : Y \to \mathbb{P}^1$ ramified exactly over $p_1, \dots, p_6$ of ramification type $(2,2,2,1)$. If we denote by $\beta : \mathcal{R}_{2,7} \to \mathcal{M}_2$ the forgetful map onto the moduli space of smooth curves of genus 2 and the intersection ${\Pr_{2,7}}^{-1}(P(f)) \cap \beta^{-1}([C])$ consists of d elements, then Lemma 3.1 implies that

$$400 = d \cdot N. \tag{3.2}$$

Let $p_1, \dots, p_6$ denote 6 points of $\mathbb{P}^1$ in general position. We want to compute the number N of isomorphism classes of degree 7 coverings ramified of ramification type $(2,2,2,1)$ exactly over the points $p_1, \dots, p_6$.

Let π_1 denote the monodromy group of $\overline{f}$ with base point $p \neq p_i$ for all $i = 1, \dots, 6$. We consider π_1 as a subgroup of the symmetric group $\mathcal{S}_7$ of degree 7 acting on the set $\{1, \dots, 7\}$. We need the following 2 trivial lemmas (which can of course be formulated for any odd prime p instead of 7).

LEMMA 3.2. *Let a and b be involutions of $\mathcal{S}_7$ such that*

$$s := ab$$

is a cycle of order 7, then the group generated by a and b is isomorphic to D_7. Conversely, any subgroup of $\mathcal{S}_7$ isomorphic to D_7 is of this type.

PROOF. For the first assertion we have to show that $asa = s^{-1}$. But

$$as = aab = b = b^{-1} = b^{-1}a^{-1}a = s^{-1}a.$$

The second assertion is obvious. □

LEMMA 3.3. *Let a and b be involutions of $\mathcal{S}_7$ such that $\langle a, b\rangle \simeq D_7$. Then a and b are products of 3 disjoint transpositions such that no transposition in a occurs in b. Equivalently a, respectively b, fixes exactly one number i, respectively j, and $i \neq j$. Each of the numbers $1, \dots, 7$ is fixed by exactly one of the 7 involutions.*

PROOF. It is easy to check that if a or b consist of less that 3 disjoint transpositions or if one transposition occurring in a occurs also in b, then $s = ab$ is not of order 7, which contradicts Lemma 3.2. For the last assertion note that if 2 involutions would fix the same number i, then all involutions and hence all elements of the group would fix i. □

We give an example, although this is not necessary for the sequel, but makes the following proposition perhaps clearer.

EXAMPLE 3.4. Consider the following involutions of $\mathcal{S}_7$ (with right action, i.e. the product is from left to right)

$$a = (12)(34)(56) \quad \text{and} \quad b = (23)(45)(67),$$

with product

$$s := ab = (1357642).$$

According to Lemma 3.2 the group $\langle a, b\rangle$ is isomorphic to D_7.

PROPOSITION 3.5. *All dihedral subgroups D_7 of $\mathcal{S}_7$ are conjugate to each other.*

PROOF. Different labelings of the set on which $\mathcal{S}_7$ acts give conjugate subgroups of $\mathcal{S}_7$. We label this set in such a way that

$$a = (12)(34)(56).$$

The labelling is unique up to the transpositions $(12), (34)$ and (56). Moreover, according to Lemma 3.3 we may choose b such that it fixes 1. The involution b cannot contain the transposition (27), since otherwise $s = ab$ would contain the cycle (172). So we remain with the following possibilities b_i for b:

$$\begin{aligned}
b_1 &= (23)(45)(67), & b_2 &= (23)(46)(57), & b_3 &= (23)(47)(56),\\
b_4 &= (24)(35)(67), & b_5 &= (24)(36)(57), & b_6 &= (24)(37)(56),\\
b_7 &= (25)(34)(67), & b_8 &= (25)(36)(47), & b_9 &= (25)(37)(46),\\
b_{10} &= (26)(34)(57), & b_{11} &= (26)(35)(47), & b_{12} &= (26)(37)(45).
\end{aligned}$$

We compute

$$s_1 = ab_1 = (1357642), \quad s_2 = ab_2 = (1367542), \quad s_3 = ab_3 = (13742),$$

$$s_4 = ab_4 = (1457632), \quad s_5 = ab_5 = (1467532), \quad s_6 = ab_3 = (14732),$$

$$s_7 = ab_7 = (15762), \quad s_8 = ab_8 = (1537462), \quad s_9 = ab_9 = (1547362),$$

$$s_{10} = ab_{10} = (16752), \quad s_{11} = ab_{11} = (1637452), \quad s_{12} = ab_{12} = (1647352).$$

So exactly the groups $G_i := \langle a, b_i \rangle$ with $i = 1, 2, 4, 5, 8, 9, 11$ and 12 are isomorphic to D_7.

We may still conjugate the subgroups with the transpositions $(12), (34)$ and (56). One checks that

$$(12)s_1(12) = s_5^6 \in G_6, \quad (34)s_1(34) = s_4 \in G_4, \quad (56)s_1(56) = s_2 \in G_2,$$

$$(12)s_8(12) = s_{12}^6 \in G_{12}, \quad (34)s_8(34) = s_9 \in G_9, \quad (56)s_8(56) = s_{11} \in G_{11}.$$

This implies that G_1, G_2, G_4 and G_5 as well as G_8, G_9, G_{11} and G_{12} are pairwise conjugate. Hence it suffices to show that G_1 is conjugate to G_8. But one easily checks that with the permutation $p := (2463)$ we have

$$pap^{-1} = a \quad \text{and} \quad pb_1p^{-1} = b_8.$$

which gives the assertion. □

PROPOSITION 3.6. *Given* 6 *points of* $\mathbb{P}^1$ *in general position. There are exactly* 400 *isomorphism classes of degree-7 coverings* $\overline{f} : Y \to \mathbb{P}^1$ *with monodromy group* D_7 *ramified of type* $(2, 2, 2, 1)$ *over each of the* 6 *points.*

PROOF. Let $p_1, \ldots, p_6 \in \mathbb{P}^1$ be 6 points in general position. According to Proposition 3.5 all subgroups isomorphic to D_7 are conjugate. For example, we may take the monodromy subgroup of the covering to be

$$\pi_1 = \langle a, b_1 \rangle = \{1, s_1, \ldots, s_1^6, a, as_1, \ldots, as_1^6\}.$$

One has to compute the number of conjugacy classes of 6 involutions c_i of type (2,2,2,1) (associating c_i to the point p_i) such that

$$c_1c_2c_3c_4c_5c_6 = 1. \tag{3.3}$$

The elements of D_7 are either an involution (an odd permutation) or an element of order 7 (an even permutation). Hence the product of 5 involutions in D_7 gives an odd permutation, so it is an involution. Thus, for any involutions $c_1, \ldots, c_5$ of π_1, $c_6 := c_1 \cdots c_5$ is an involution satisfying (3.3). Since any 2 involutions of π_1 generate the group, only 7 of the corresponding coverings are not connected, namely those with $c_1 = \cdots = c_6$. This gives $7^5 - 7$ coverings. Two 6-tuples of involutions give isomorphic coverings if and only if they are conjugate.

In order to compute the set of conjugate classes of 6-tuples of involutions, recall that the only transitive subgroups of $\mathcal{S}_7$ which are not contained in $\mathcal{A}_7$ are subgroups isomorphic to

- the dihedral group D_7,
- the group $L_7 = AGL_1(\mathbb{F}_7)$ of affine transformations of the line with 7 points,
- $\mathcal{S}_7$ itself.

(For the convenience of the reader we give a proof of this statement: Let G be a proper subgroup of $\mathcal{S}_7$ of this type. Clearly G is a soluble group, since it is not contained in $\mathcal{A}_7$. It is primitive, being a permutation group of degree 7. Hence, according to [**12**, 7.2.7], it is a subgroup of L_7. Since D_7 is the only transitive subgroup of L_7, this gives the statement.) The group D_7 is not normal in $\mathcal{S}_7$, but it is normal in L_7. Since on the other hand L_7 is isomorphic to the group of outer automorphisms of D_7, two 6-tuples of involutions give isomorphic coverings if and only if they are conjugate under an element of L_7. Since L_7 is of order 42, we finally get

$$\frac{7^5 - 7}{42} = 400$$

isomorphism classes of degree 7 coverings $Y \to \mathbb{P}^1$ of ramification type $(2,2,2,1)$. □

Together with (3.2) we get

COROLLARY 3.7. *Let $\overline{f} : Y \to \mathbb{P}^1$ be a degree 7 covering with monodromy group D_7 ramified of type $(2,2,2,1)$ over 6 general points. There is exactly one cyclic étale cover $f : \widetilde{C} \to C$ such that the diagram (3.1) is commutative. In particular, $P(f) \simeq JY \times JY$ as polarized abelian varieties of type $(1,\dots,1,7)$.*

4. Proof of the main theorem

Let $\mathcal{C}_3$ denote the locus in $\mathcal{M}_3$ of genus 3 curves admitting a degree 7 covering $\overline{f} : Y \to \mathbb{P}^1$ with ramification type $(2,2,2,1)$ over 6 general points in $\mathbb{P}^1$. It is an irreducible variety of dimension 3 since it is an image of the moduli space of pairs $(Y, \overline{f})$, which according to Lemma 3.1 are in bijection with the elements in $\mathcal{R}_{2,7}$.

THEOREM 4.1. *The elements of a generic fiber of the Prym map $\mathrm{Pr}_{2,7} : \mathcal{R}_{2,7} \to \mathcal{B}_D$ are in bijection with the set of degree 7 coverings $\overline{f} : Y \to \mathbb{P}^1$ with ramification type $(2,2,2,1)$ over 6 general points that a curve $Y \in \mathcal{C}_3$ admits.*

PROOF. By Corollary 3.7 the elements in the fiber of $P(f) \in \mathcal{B}_D$ are necessarily in different fibers of the forgetful map $\beta : \mathcal{R}_{2,7} \to \mathcal{M}_2$. Together with uniqueness of the decomposition $P(f) \simeq JY \times JY$ as polarized abelian varieties (see Theorem 2.6) this implies that the covering $f : \widetilde{C} \to C$ is completely determined by the pair $(Y, \overline{f})$, since the branch points of $\overline{f}$ determine the genus 2 curve C. So the elements in $\mathcal{R}_{2,7}$ with Prym variety isomorphic to $JY \times JY$ are in bijection with the coverings $\overline{f} : Y \to \mathbb{P}^1$ with ramification type $(2,2,2,1)$ for a fixed curve $Y \in \mathcal{C}_3$. □

It has been shown in [**8**, Theorem 1.1] that the degree of the Prym map $\mathrm{Pr}_{2,7}$ is 10, so we get as immediate corollary

COROLLARY 4.2. *There are at most 10 coverings $\overline{f} : Y \to \mathbb{P}^1$ of degree 7 with ramification type $(2,2,2,1)$ for a curve $Y \in \mathcal{C}_3$.*

References

[1] Arnaud Beauville, *Prym varieties and the Schottky problem*, Invent. Math. **41** (1977), no. 2, 149–196. MR0572974

[2] Christina Birkenhake and Herbert Lange, *Complex abelian varieties*, 2nd ed., Grundlehren der Mathematischen Wissenschaften [Fundamental Principles of Mathematical Sciences], vol. 302, Springer-Verlag, Berlin, 2004. MR2062673

[3] Ch. Birkenhake and H. Lange, *Moduli spaces of abelian surfaces with isogeny*, Geometry and analysis (Bombay, 1992), Tata Inst. Fund. Res., Bombay, 1995, pp. 225–243. MR1351509

[4] Ron Donagi, *The fibers of the Prym map*, Curves, Jacobians, and abelian varieties (Amherst, MA, 1990), Contemp. Math., vol. 136, Amer. Math. Soc., Providence, RI, 1992, pp. 55–125. MR1188194

[5] Ron Donagi and Roy Campbell Smith, *The structure of the Prym map*, Acta Math. **146** (1981), no. 1-2, 25–102. MR594627

[6] Carel Faber, *Prym varieties of triple cyclic covers*, Math. Z. **199** (1988), no. 1, 61–79. MR954752

[7] Herbert Lange and Angela Ortega, *Prym varieties of triple coverings*, Int. Math. Res. Not. IMRN **22** (2011), 5045–5075. MR2854722

[8] Herbert Lange and Angela Ortega, *The Prym map of degree-7 cyclic coverings*, Algebra Number Theory **10** (2016), no. 4, 771–801. MR3519095

[9] David Mumford, *Prym varieties. I*, Contributions to analysis (a collection of papers dedicated to Lipman Bers), Academic Press, New York, 1974, pp. 325–350. MR0379510

[10] Angela Ortega, *Variétés de Prym associées aux revêtements n-cycliques d'une courbe hyperelliptique* (French), Math. Z. **245** (2003), no. 1, 97–103. MR2023955

[11] Herbert Popp, *Fundamentalgruppen algebraischer Mannigfaltigkeiten* (German), Lecture Notes in Mathematics, Vol. 176, Springer-Verlag, Berlin-New York, 1970. MR0277540

[12] Derek J. S. Robinson, *A course in the theory of groups*, 2nd ed., Graduate Texts in Mathematics, vol. 80, Springer-Verlag, New York, 1996. MR1357169

Department Mathematik der Universität Erlangen-Nürnberg, Cauerstr. 11, 91058 Erlangen, Germany

Email address: `lange@math.fau.de`

Institut für Mathematik, Humboldt Universität zu Berlin, Unter den Linden 6, D-10099 Berlin, Germany

Email address: `ortega@math.hu-berlin.de`

Contemporary Mathematics
Volume **709**, 2018
http://dx.doi.org/10.1090/conm/709/14294

Extremal higher codimension cycles of the space of complete conics

César Lozano Huerta

Abstract. Let M denote the space of complete conics. We compute the cone of effective and numerically effective k-cycles of M, $\mathrm{Eff}_k(M)$ and $\mathrm{Nef}_k(M)$, respectively. In addition, we compute the Białynicki-Birula cell-decomposition of M with respect to a $\mathbb{C}^*$-action and compare the cone generated by the closure of these cells to the cones $\mathrm{Eff}_k(M)$ and $\mathrm{Nef}_k(M)$.

Introduction

The study of how a subvariety sits and moves inside an ambient algebraic variety has driven research in algebraic geometry for a long time. This study finds a robust and well-developed theory when dealing with subvarieties of codimension 1, which from the theoretical viewpoint is well-understood. For example, studying points on an algebraic curve leads to theorems such as the Riemann-Roch Theorem and the Abel-Jacobi Theorem. A sensible next step in understanding the extrinsic geometry of subvarieties is to study subvarieties of codimension 2 and higher. This situation is not well-understood and is the situation we will study. We work over the field of complex numbers throughout.

The set of all subvarieties contained in an algebraic variety is unmanageable. Thus, one seeks to impose an equivalence relation on them, or rather on the formal linear combinations of them, so we study the set of equivalence classes which may be more manageable. These formal sums are called cycles, and Severi [**13**] in a collection of papers set up and investigated the concept of rational equivalence on them. Although this is a generalization inspired by the rational equivalence of divisors on curves, it exhibits distinct behavior. For example, Mumford [**11**] showed that the group of 0-cycles, up to rational equivalence, on a surface does not necessarily have the properties of a finite-dimensional space. We will study the space of complete conics M and the spaces of its rational-equivalent cycles $A_k(M) \otimes \mathbb{Q}$; these are finite-dimensional vector spaces.

The space of complete conics is a compactification of the family of smooth conics on the plane $\mathbb{P}^2$. It was introduced by Chasles [**2**] in 1864 in order to solve questions

2010 *Mathematics Subject Classification.* Primary 14C25, 14C99; Secondary 14E30.
Key words and phrases. Algebraic geometry, complete conics, higher codimension cycles.
During the preparation of this document the author was partly supported by CONACYT: CB 2015/253061. The author is currently a CONACYT-Research Fellow at UNAM, ID: 71842.

in enumerative geometry and is defined as follows: the space of complete conics is the closure of the set of pairs $M = \overline{\{(C, C^*)\}} \subset \mathbb{P}^5 \times \mathbb{P}^{5*}$, where C is a smooth conic and C^* its dual. This space is a smooth variety of dimension 5, and has been used to study enumerative questions such as the number of conics tangent to five fixed conics in general position; which is 3264 and has a long history in algebraic geometry [**10**]. The purpose of this note is to show that the enumerative geometry of conics that M allows one to study dictates the structure of certain cones contained in the space of cycles $A_k(M)$: the cones of effective cycles, $\mathrm{Eff}_k(M)$.

We are primarily interested in understanding subvarieties of M of dimension k, so we will focus on the cone generated by non-negative linear combinations with rational coefficients of subvarieties of dimension k in the space of k-cycles $A_k(M) \otimes \mathbb{Q}$. This is the cone of effective k-cycles, $\mathrm{Eff}_k(M)$. In general, these cones are very difficult to compute and only few examples are known; many of them very recently [**4, 3**]. We will explicitly compute these cones for the space M.

These cones are well-known in dimension 1 and 4. Indeed, from the definition of M, there are 4 divisors we can investigate: $H_i = \pi_i^*(h)$ and the strict transform $E_i = \pi_i^{-1}(\Delta)$, where π_i $(i = 1, 2)$ denotes the projection map to the first or second factor. Here, h denotes the hyperplane section, and Δ denotes the locus of singular conics *i.e.*, a discriminant variety. The cone of effective divisors is generated by the classes $\mathrm{Eff}_4(M) = \langle E_1, E_2 \rangle$, and the cone of numerically effective divisors, the nef cone, is generated by the classes $\overline{\mathrm{Eff}}_1(M)^\vee = \mathrm{Nef}^1(M) = \langle H_1, H_2 \rangle$. The divisor E_2 parametrizes reducible conics, and E_1 double lines with two marked-points. A generic point on H_1 parameterizes a smooth conic which passes through a fixed point, and a generic point on H_2 parametrizes a smooth conic tangent to a fixed line.

The main results of this note, Theorem 3.1 and Theorem 4.1, extend the description of the previous paragraph to dimensions 2 and 3, *i.e.*, we describe the cones $\mathrm{Eff}_k(M)$ for $k = 2, 3$. In addition, one can read off the geometric interpretation of the generators of these cones in terms of families of conics. To simplify the language, H_iE_i denotes the intersection of the subvarieties H_i and E_i.

THEOREM. *The vector spaces $A_k(M) \otimes \mathbb{Q}$ for $k = 2, 3$, both have dimension* 3. *The respective cones of effective cycles are generated by the following classes*

$$\mathrm{Eff}_2(M) = \langle H_1^2E_1,\ H_2^2E_2,\ H_1E_1E_2,\ H_2E_1E_2 \rangle,$$
$$\mathrm{Eff}_3(M) = \langle H_1E_1, H_2E_2, E_1E_2 \rangle.$$

An important part of the proof is to reduce the computation of the effective cone $\mathrm{Eff}_k(M)$ to the computation of the effective cone of two subvarieties: $\mathrm{Eff}_k(E_1)$ and $\mathrm{Eff}_k(E_1E_2)$. We are able to do this because there is an action of $SL_3\mathbb{C}$ on M, and E_1 as well as E_1E_2 are the closure of SL_3-orbits. Our technique may be applied in the more general context in which the orbits of a group action behave well; for example "wonderful compactifications" [**5**]. Since the action of $SL_3\mathbb{C}$ is ultimately the key ingredient in order to compute the effective cones, one may ask whether the action of a smaller group would yield the same results. For example:

The space M inherits from $\mathbb{P}^2$ an action of the torus $\mathbb{C}^*$ which has a finite number of fixed points. Thus, it induces a Białynicki-Birula decomposition of M into affine

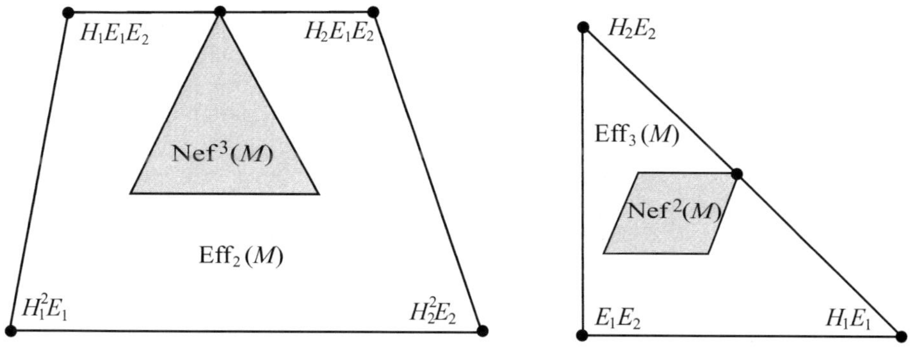

FIGURE 1. Cross-section of the cones $\mathrm{Eff}_k(M)$, and their duals $\mathrm{Nef}^k(M)$.

cells [**1**]. We call these cells the BB-cells of M. By considering the closure of the BB-cells of dimension k, we get effective k-cycles on M and it is natural to ask, do they generate the cone of effective cycles of dimension k, $\mathrm{Eff}_k(M)$? In general, we answer in the negative by computing the number of BB-cells of dimension 2 and the cone $\mathrm{Eff}_2(M)$. However, the BB-cells of M of dimension 4 do generate the cone of effective divisors $\mathrm{Eff}_4(M)$. In general, the precise relation between the cone generated by the BB-cells of a $\mathbb{C}^*$-variety X and the cones of effective cycles $\mathrm{Eff}_k(X)$ remains mysterious.

1. Preliminaries

In this section, we recall definitions and properties of M. We refer the reader to [**6**] for a more general treatment. A *cycle* on M of dimension k is a formal sum of k-dimensional subvarieties of M with rational coefficients. Such a cycle is called *effective* if all the coefficients of this formal sum are nonnegative. Two cycles X and Y of dimension k are numerically equivalent if $X \cdot \alpha = Y \cdot \alpha$ for all irreducible subvarieties of codimension k. Here, $\cdot$ denotes the intersection product. Let $N_k(M)$ (respectively, $N^k(M)$) denote the finite dimensional $\mathbb{Q}$-vector space of cycles of dimension k (respectively, codimension k) up to numerical equivalence. The cone in $N_k(M)$ generated by the classes of effective cycles of dimension k is called *effective cone* and is denoted by $\mathrm{Eff}_k(M)$.

A cycle Z of dimension k is numerically effective, or simply *nef*, if $Z \cdot \alpha \geq 0$ for all irreducible subvarieties $\alpha \subset M$ of codimension k. One defines the *nef cone* of cycle classes of dimension k, $\mathrm{Nef}_k(M)$, as the cone generated by the classes in $N_k(M)$ of nef cycles. Similarly, one defines $\mathrm{Nef}^k(M)$, the cone of nef cycle classes of codimension k. Note that in general the dual of the cone Nef^k is the closure of the effective cone $\overline{\mathrm{Eff}}_k$. These convex cones are effectively described by extremal rays. Recall that a ray R is said to be an *extremal ray* if for any $D \in R$ with $D = D_1 + D_2$, where D_1, D_2 are in the cone, then $D_1, D_2 \in R$. If an extremal ray is spanned by the class of an effective cycle Z, we say that Z is an *extremal effective cycle.* In general, Eff_k is not equal to its closure. In our case, we have that $\overline{\mathrm{Eff}}_k(M) = \mathrm{Eff}_k(M)$ as a consequence of the results of the present paper.

The space of complete conics M admits the following two alternative descriptions. First, $M \cong \overline{M}_{0,0}(\mathbb{P}^2, 2)$ is isomorphic to a Kontsevich moduli space of stable maps. This description of M often facilitates to perform intersection theory, but we will make little use of this isomorphism. On the other hand, it is important for us that M is isomorphic to the blowup of $\mathbb{P}^5$ along the Veronese surface $S \subset \mathbb{P}^5$, [**8**]. The exceptional divisor of this blowup coincides with E_1. In fact, from the definition of M as the closure of the set of pairs $\{(C, C^*)\}$, the projection into the first factor is the blowup map. Furthermore, this description of M allows one to compute $A^k(M)$ for $k = 1, 2, 3$, and see that rational equivalence and numerical equivalence of cycles coincide on M. In fact, rational and numerical equivalence coincide for a large class of varieties called spherical varieties [**12**]; M is a variety of this type. Therefore, we may use freely $N_k(M)$ or $A_k(M) \otimes \mathbb{Q}$.

The space M inherits an action of SL_3 from $\mathbb{P}^2$. This action induces the following stratification into SL_3-orbits [**5**]

$$M = U \cup (E_1^\circ \cup E_2^\circ) \cup E_1E_2,$$

where U is an open dense subset, and E_1° and E_2° are SL_3-orbits of codimension 1. The unique closed orbit is E_1E_2.

The characteristic numbers of M such as $H_2^5 = 1$ (the number of conics tangent to five fixed lines in general position) have a long history in algebraic geometry. We make use of such numbers in order to understand the geometry of the cones $\mathrm{Eff}_k(M)$ and $\mathrm{Nef}_k(M)$. We refer the reader to [**7**] for a proof of the following

LEMMA 1.1. *Let M be the space of complete conics. The following intersection numbers hold,*

$$H_1^5 = H_2^5 = 1, \quad H_1^4H_2 = H_1H_2^4 = 2, \quad H_1^2H_2^3 = H_1^3H_2^2 = 4.$$

Using this lemma, it is straightforward to obtain that $H_2 = 2H_1 - E_1$ and $H_1 = 2H_2 - E_2$. There is a $\mathbb{Z}/2$-symmetry in the previous lemma induced by the Gauss map $C \mapsto C^*$, where C^* denotes the dual conic of C. We make use of this symmetry in order to simplify our computations.

2. Białynicki-Birula decomposition of M

In this section, we compute the dimension of the affine cells of the Białynicki-Birula decomposition of M following [**1**]. This decomposition is induced by a $\mathbb{C}^*$-action on M inherited from the plane $\mathbb{P}^2$. Let us state, in our context, the theorem we will use.

THEOREM. [**1**] *Let X be a smooth projective variety. Suppose there is an action of $\mathbb{C}^*$ on X. If this action has a finite number of isolated fixed points $\{p_1, \dots, p_n\}$, then*

- *There exists a decomposition $X = \bigcup_i C_{p_i}$, where each C_{p_i} is an affine variety containing a unique p_i, called center.*
- *The tangent space at any p_i, $T_{p_i}X$, becomes a $\mathbb{C}^*$-module. Hence, there is a decomposition*

$$T_{p_i}X = T^+ \oplus T^{-1},$$

where T^{-1} denotes the subspace upon which the $\mathbb{C}^*$-action has negative weights, and $C_{p_i} \cong T^{-1}$ as affine varieties.

Since $M \cong \mathrm{Bl}_S(\mathbb{P}^5)$ is isomorphic to the blowup of $\mathbb{P}^5$ along the Veronese surface S, we will compute the dimension of the BB-cells of M by first computing the BB-cells of $\mathbb{P}^5$. Let us denote by $[a : \ldots : f]$ the homogeneous coordinates for the linear system $|\mathcal{O}_{\mathbb{P}^2}(2)| \cong \mathbb{P}^5$. There is an induced action of $\mathbb{C}^* \times \mathbb{C}^*$ on $|O_{\mathbb{P}^2}(2)|$ given by $[at_0^2 : bt_0t_1 : ct_1^2 : dt_0 : et_1 : f]$. Let T be the image of $\mathbb{C}^*$ in $\mathbb{C}^* \times \mathbb{C}^*$ under the embedding $t \mapsto (t, t^k)$, $k \gg 0$. We call the action of this embedded torus T-action. This action of T on $\mathbb{P}^5$ has six isolated fixed points, $p_i = [0 : \ldots : 1 : \ldots : 0] \in \mathbb{P}^5$, which implies that it induces a Białynicki-Birula decomposition of $\mathbb{P}^5$. According to the Theorem above, the dimension of each BB-cell is the number of negative weights of the T-action on each tangent space $T_{p_i}\mathbb{P}^5$. The following table shows the dimension of the BB-cells, and their respective centers. It also shows $\mathbb{C}^* \times \mathbb{C}^*$ acting on each tangent space $T_{p_i}\mathbb{P}^5$ by a diagonal matrix, which we denote by diag.

(2.1)

center	cell-dimension	$\mathbb{C}^* \times \mathbb{C}^*$-action on $T_{p_i}\mathbb{P}^5$
p_0	2	$\mathrm{diag}(t_0^{-1}t_1, t_0^{-2}t_1^2, t_0^{-1}, t_1t_0^{-2}, t_0^{-2})$
p_1	4	$\mathrm{diag}(t_0t_1^{-1}, t_0^{-1}t_1, t_1^{-1}, t_0^{-1}, t_0^{-1}t_1^{-1})$
p_2	5	$\mathrm{diag}(t_0^2t_1^{-2}, t_0t_1^{-1}, t_0t_1^{-2}, t_1^{-1}, t_1^{-2})$
p_3	1	$\mathrm{diag}(t_0, t_1, t_1^2t_0^{-1}, t_0^{-1}t_1, t_0^{-1})$
p_4	3	$\mathrm{diag}(t_0^2t_1^{-1}, t_0, t_1, t_0t_1^{-1}, t_1^{-1})$
p_5	0	$\mathrm{diag}(t_0^2, t_0t_1, t_1^2, t_0, t_1)$

In order to describe the BB-decomposition of the space of complete conics $M \cong \mathrm{Bl}_S(\mathbb{P}^5)$, let us first determine the fixed points of the action of T on M. Since the surface S is invariant under the action of T on $\mathbb{P}^5$, then we may write the action of T on M by analyzing how it acts on the normal bundle $N_{S/\mathbb{P}^5}$.

The surface S is the image of the map $\nu : [x, y, z] \mapsto [x^2 : xy : y^2 : xz : yz : z^2]$. Three T-fixed points of $\mathbb{P}^5$ lie on the Veronese surface $S = \nu(\mathbb{P}^2) \subset \mathbb{P}^5$, namely p_0, p_2, p_5. It suffices to analyze the action of T on the fibers $\pi^{-1}(p_i)$, $i = 0, 2, 5$, where $\pi : M \to \mathbb{P}^5$ denotes the blowup map. Here, we only examine the case for the fiber $\pi^{-1}(p_0)$, as the other two cases are similar. The point p_0 is the image of $[1 : 0 : 0]$ under ν, hence looking at the affine chart $p_0 \neq 0$ of $\mathbb{P}^5$, the derivative

$$d\nu_{|(0,0)} = \begin{pmatrix} 1 & 0 & 0 & 0 & 0 \\ 0 & 0 & 1 & 0 & 0 \end{pmatrix}$$

tells us how the tangent space $T_{p_0}S$ sits inside the tangent space $T_{p_0}\mathbb{P}^5$. Since we know that $\mathbb{C}^* \times \mathbb{C}^*$ acts on $T_{p_0}\mathbb{P}^5$ by the diagonal matrix indicated in (2.1), we conclude that $\mathbb{C}^* \times \mathbb{C}^*$ acts on both $T_{p_0}S$, and on the fiber over p_0 of the normal bundle $N_{S/\mathbb{P}^5}$, by the following matrices, respectively

$$\begin{pmatrix} t_0^{-1}t_1 & \\ & t_0^{-1} \end{pmatrix}, \quad \begin{pmatrix} t_0^{-2}t_1^2 & & \\ & t_1t_0^{-2} & \\ & & t_0^{-2} \end{pmatrix}. \tag{2.2}$$

On the fiber $\pi^{-1}(p_0)$, we have three T-fixed points; let us denote them by p_0', p_0'', p_0'''. Similarly, we have six more points p_2', p_2'', p_2''', and p_5', p_5'', p_5''', from which we conclude that there are 12 points on M fixed by T; namely $p_0', p_0'', \ldots, p_5''', p_1, p_3, p_4$.

The table 2.1 lists the dimension of the BB-cells of M with centers p_1, p_3, p_4. In order to complete the computation of the dimension of the BB-cells of M, we now compute the dimension of the BB-cells whose centers are $p_0', p_0'', \ldots, p_5'''$.

Let us compute the dimension of the BB-cell of M whose center is p_0'. If E stands for the exceptional divisor of the blowup $\pi : M \to \mathbb{P}^5$, then we have that $T_{p_0'}M = T_{p_0'}E \oplus N_{E/M}(p_0')$, and this decomposition is compatible with the action of T. The space $T_{p_0'}E$ can be further decomposed as $T_{p_0'}E = T_{p_0'}S \oplus T_{p_0'}F$, where $T_{p_0'}S$ denotes tangent space to the Veronese surface $S = \nu(\mathbb{P}^2)$, with the fiber $F = \pi^{-1}(p_0)$. This decomposition is also compatible with the action of T. Therefore, we have

$$T_{p_0'}M = T_{p_0'}E \oplus N_{E/M}(p_0') = T_{p_0'}S \oplus T_{p_0'}F \oplus N_{E/M}(p_0').$$

In order to compute the dimension of the BB-cell at p_0', we count the negative weights of the action of T on each of the factor above. In fact, the matrices in (2.2) reveal that T acts with one negative weight on $T_{p_0'}S$ and two negative weights on $T_{p_0'}F$. Since $N_{E/M} \cong \mathcal{O}_E(-1)$ is the tautological bundle over E, we have that the action of $\mathbb{C}^* \times \mathbb{C}^*$ on the fiber $\mathcal{O}_E(-1)_{p_0'}$ is given by the coordinate of p_0', which is $t_0^{-2}t_1^2$, and has positive T-weight. We conclude that the dimension of the BB-cell whose center is p_0' is $1 + 2 + 0 = 3$.

The rightmost column of the following table lists the dimension of the BB-cells of M induced by T. The centers of the BB-cells are listed in the leftmost column. We also list the negative weights of the T-action on each of the factors $T_qS, T_qF, N_{E/M}(q)$.

point	T_qS	T_qF	$N_{E/M}(q)$	cell-dimension
p_0'	1	2	0	3
p_0''	1	1	0	2
p_0'''	1	0	1	2
p_2'	2	2	1	5
p_2''	2	1	1	4
p_2'''	2	0	1	3
p_5'	0	0	0	0
p_5''	0	1	0	1
p_5'''	0	2	0	2

The BB-cells with centers at p_1, p_3, p_4 have dimensions $4, 1, 3$, respectively. Hence, there are three BB-cells of M of dimensions 3 and 2. Also, there are two BB-cells of dimensions 1 and 4.

3. Cycles of M of dimension 3

THEOREM 3.1. *The space $A_3(M) \otimes \mathbb{Q}$ has dimension 3. The cone of effective 3-cycles is generated by the following classes*

$$\mathrm{Eff}_3(M) = \langle H_1E_1, H_2E_2, E_1E_2 \rangle.$$

PROOF. The dimension of $A_3(M) \otimes \mathbb{Q}$ is 3 since there are three BB-cells of dimension 3 [**6**].

In order to describe the cone of effective 3-cycles, we make use of the following strategy. We know that $\langle H_1E_1, H_2E_2, E_1E_2\rangle \subset \mathrm{Eff}_3(M)$. We want to show that the opposite containment holds, which is equivalent to showing the following inclusion of the dual cones $\langle H_1E_1, H_2E_2, E_1E_2\rangle^\vee \subset \mathrm{Eff}_3(M)^\vee$. By definition, we have $\overline{\mathrm{Eff}}_3(M)^\vee = \mathrm{Nef}^3(M)$. Hence, we get the desired inclusion of dual cones as long as the classes in $\langle H_1E_1, H_2E_2, E_1E_2\rangle^\vee$ are nef. Nefness is what we prove next.

By definition, the dual cone $\langle H_1E_1, H_2E_2, E_1E_2\rangle^\vee$ consists of all the classes $\alpha \in A_2(M)$ such that $\alpha\cdot\beta \geq 0$, where $\beta \in \langle H_1E_1, H_2E_2, E_1E_2\rangle$. In terms of a basis, this definition reads as follows. We need $a, b, c \in \mathbb{Q}$ such that $(aH_1^2E_1 + bH_2^2E_2 + cH_1^3)(lH_1E_1 + rH_2E_2 + sE_1E_2) \geq 0$, where $r, l, s \geq 0$, and $\{H_1^2E_1, H_2^2E_2, H_1^3\}$ is a basis for $A_2(M) \otimes \mathbb{Q}$ (the fact that this space has dimension 3 is proved in Theorem 4.1). We split up this inequality into three which, using Lemma 1.1 and $E_1 = 2H_1 - H_2$ and $E_2 = 2H_2 - H_1$, yields

$$\begin{aligned}
H_1E_1.\alpha &= H_1E_1(aH_1^2E_1 + bH_2^2E_2 + cH_1^3) \\
&= 8b \ \geq 0, \\
H_2E_2.\alpha &= H_2E_2(aH_1^2E_1 + bH_2^2E_2 + cH_1^3) \\
&= 8a + 6c \geq 0, \\
E_1E_2.\alpha &= E_1E_2(aH_1^2E_1 + bH_2^2E_2 + cH_1^3) \\
&= -8a - 8b \geq 0\ .
\end{aligned}$$

From these three inequalities, it follows that the cone $\langle H_1E_1, H_2E_2, E_1E_2\rangle^\vee$ is generated by the following three classes,

$$\alpha_1 = 4H_1^3 - 3H_1^2E_1, \quad \alpha_2 = -3H_1^2E_1 + 3H_2^2E_2 + 4H_1^3, \quad \alpha_3 = H_1^3.$$

Since $A_2(M)\otimes\mathbb{Q}$ has dimension 3, the linear span of $\{H_1^3, H_2^3, H_1^2H_2, H_2^2H_1\}$ must have at least one linear relation. Indeed, set $xH_1^3 + yH_2^3 + zH_1^2H_2 + wH_2^2H_1 = 0$, for $x, y, z, w \in \mathbb{Q}$. By intersecting this linear combination with the classes H_1^2, H_1H_2 and E_1E_2, we find linear relations among x, y, z, w. Such relations yield $2H_1^3 - 3H_1^2H_2 + 3H_1H_2^2 - 2H_2^3 = 0$, which implies that $2\alpha_1 = (H_1 + H_2)E_1E_2$ and $\alpha_2 = 4H_2^3$.

Since H_1^3, H_2^3 are nef classes, the result follows if the class $\tau = (H_1 + H_2)E_1E_2$ is also nef. In other words, $\mathrm{Eff}_3(M) = \langle H_1E_1, H_2E_2, E_1E_2\rangle$ if and only if the class τ is a nef class. Nefness is proved in the next Lemma. □

LEMMA 3.2. *The class* $\tau = (H_1 + H_2)E_1E_2 \in A_2(M)$ *is nef.*

PROOF. Throughout this proof Z denotes an irreducible threefold in M. We wish to show that $Z.\tau \geq 0$ for any Z.

The space M is stratified into $SL_3\mathbb{C}$-orbits and we can differentiate the irreducible threefolds in M based on how they intersect these strata. We will prove this lemma by considering all the possibilities for these intersections and showing that in each of them, we have that $Z.\tau \geq 0$. The first case is the following. If Z intersects transversally all the strata, then the intersection $Z \cap E_1E_2$ is an irreducible curve, which implies that $Z.\tau \geq 0$, as τ is an ample divisor in E_1E_2.

Let us analyze the cases in which the intersection of Z with the SL_3-strata may not be transversal. Since $\tau \subset E_1E_2$, it suffices to analyze the dimensions of the intersections $Z \cap E_1$ and $Z \cap E_1E_2$ listed below. The rows of the following table show the possibilities we are going to examine.

case	dim $Z \cap E_1$	dim $Z \cap E_1E_2$
I	2	2,1
II	3	3, 2

(I) Suppose dim $Z \cap E_1 = 2$, and dim $Z \cap E_1E_2 = 1$. Since $H_1 + H_2$ is the hyperplane section of the embedding $M \subset \mathbb{P}^9 \times \mathbb{P}^{9*}$, then τ is a very ample divisor in E_1E_2. By Kleiman's Transversality Theorem **[9]** the translate $Z^g \cap E_1E_2$, where $g \in U \subset SL_3$ and U is open dense, is an irreducible curve. Then, $Z^g.\tau > 0$ which implies that $Z.\tau \geq 0$.

Suppose dim $Z \cap E_1 = 2$ and dim $Z \cap E_1E_2 = 2$. This implies the intersection $Z.\tau$ may be considered as taking place in E_1. In other words, there is a cycle $\tau' \in A_2(E_1)$ such that $\iota_*(\tau') = \tau$, where $\iota : E_1 \to M$ stands for the inclusion map and $Z.\tau = Z_{|E_1}.\tau'$. Thus, it suffices to show that τ' is a nef class in $A_2(E_1) \otimes \mathbb{Q}$. This is proved in Lemma 3.4.

(II) Suppose dim $Z \cap E_1 = 3$, and dim $Z \cap E_2 = 3$. This, implies that $Z \subset E_1E_2$. Since Z and E_1E_2 are both irreducible subvarieties, then $[Z] = \beta E_1E_2$ for some $\beta > 0$. Hence, from Lemma 1.1, it follows that $Z.\tau \geq 0$.

Secondly, suppose dim $Z \cap E_1 = 3$ and dim $Z \cap E_2 = 2$. Since Z is irreducible, then it is a divisor in E_1, and thus $[Z]$ can be written as a nonnegative linear combination of the generators of $\mathrm{Eff}_4(E_1)$. Namely, $[Z] = \alpha H_1E_1 + \beta E_1E_2$, where $\beta, \alpha \geq 0$ by Remark 3.3. Consequently, using Lemma 1.1 and the relations $E_1 = 2H_1 - H_2$ and $E_2 = 2H_2 - H_1$, we get

$$\begin{aligned} Z.\tau &= (\alpha H_1E_1 + \beta E_1E_2).\tau \\ &= (\alpha H_1E_1 + \beta E_1E_2).(H_1 + H_2)E_1E_2 \\ &= (\alpha H_1E_1 + \beta E_1E_2).(H_1E_1E_2 + H_2E_1E_2) \\ &= \alpha E_1E_2(H_1E_1.H_1 + H_1E_1.H_2) + \beta(E_1E_2.H_1E_1E_2 + E_1E_2.H_2E_1E_2) \\ &= (\beta E_1E_2)(H_1 + H_2)E_1E_2 \geq 0. \end{aligned}$$

□

The divisor E_1, which parametrizes double lines with two marked-points, is isomorphic to the Hilbert scheme of 2 points on the plane; denoted by $\mathbb{P}^{2[2]}$. Indeed, the morphism $f : E_1 \to \mathbb{P}^{2[2]}$ that maps a double line with two marked-points to the subscheme of $\mathbb{P}^2$ supported on such marked-points induces an isomorphism by Zariski Main Theorem. Let ch be the Hilbert-Chow morphism

$$ch : \mathbb{P}^{2[2]} \longrightarrow \mathbb{P}^{2(2)},$$

where $\mathbb{P}^{2(2)}$ is the symmetric product. Let us denote by $B = \mathrm{exc}(ch)$ the exceptional locus of the morphism ch.

REMARK 3.3. *The cone of effective divisors of* $\mathbb{P}^{2[2]}$ *is generated by the following divisor classes* $\mathrm{Eff}^1(\mathbb{P}^{2[2]}) = \langle B, F \rangle$, *where* F *is the divisor which induces the morphism* $g : E_1 \to \mathbb{P}^{2*}$ *which maps a subscheme* $Z \in \mathbb{P}^{2[2]}$ *to the unique line that* *contains it. Let $\iota : E_1 \to M$ be inclusion map. Observe that the induced morphism on cycles $\iota_* : A_2(E_1) \otimes \mathbb{Q} \to A_2(M) \otimes \mathbb{Q}$ is injective (Lemma 4.3), and that $\iota_* B \sim E_1E_2$ as well as $\iota_* F \sim H_1E_1$.*

We may interpret the morphisms ch and g in terms of the geometry of M. Let$\pi_i : M \to \mathbb{P}^5$, $(i = 1, 2)$, be the projection maps from the definition of M. Observe that $\pi_{1|E_1} : E_1 \to \mathbb{P}^{2*}$ is the morphism g, and $\pi_{2|E_1} : E_1 \to \Delta$ is the Hilbert-Chow morphism ch.

Let us define two cycles of dimension 2 in E_1. Let ξ be the locus of E_1 ($\cong \mathbb{P}^{2[2]}$) of non-reduced subschemes of length 2 whose support lies on a fixed line. Let σ be the locus of E_1 of non-reduced subschemes of length 2 whose unique lines that contains it belongs to a fixed pencil of lines. These two cycles are further analyzed in Lemma 4.3. Recall that $\tau = (H_1 + H_2)E_1E_2$ and that $\iota : E_1 \to M$ denotes the inclusion map.

LEMMA 3.4. *The class $[\tau'] = [\sigma + \xi] \in A_2(E_1) \otimes \mathbb{Q}$ is nef. Moreover, $\iota_*\tau' \sim \tau$.*

PROOF. Notice that the classes $[\sigma] \sim \iota^* H_1E_2$ and $[\xi] \sim \iota^* H_2E_2$ which implies that $\iota_*(\sigma + \xi) \sim \tau$. See Lemma 4.3.

In order to show nefness of the class τ', let us apply the same argument as in Lemma 3.2. We are going to omit the details of the intersection theory of E_1, but the reader must keep in mind that the enumerative geometry of the elements of E_1 (for example, the appropriate analog of Lemma 1.1), is necessary in order to carry out the following analysis.

We want to show that $\tau'.Z \geq 0$ for every irreducible surface $Z \subset E_1$. Observe that the decomposition of E_1 into SL_3-orbits is the following $E_1 = E_1^\circ \cup B$, where B parametrizes double lines with a unique marked-point. Now, we analyze the possibilities for the surface Z to intersect these strata. The first case is the following. Let us suppose that $Z \subset E_1$ intersects all the strata transversally. This implies that $Z \cap B$ is an irreducible curve, which implies that $Z.\tau' \geq 0$ as τ' is an ample divisor in B.

Let us suppose that $\dim Z \cap B = 2$. In this case, $Z \subset B$, which implies that $[Z] = a\sigma + b\xi$, for $a, b \geq 0$ rational numbers (Lemma 4.3). Therefore

$$\begin{aligned} Z.\tau' &= (a\sigma + b\xi).\tau' \\ &= (a\sigma + b\xi).(\sigma + \xi) = 0. \end{aligned}$$

□

COROLLARY 3.5. *The cone of numerically effective 2-cycles is generated by the following classes* $\mathrm{Nef}^3(M) = \langle H_1^3, H_2^3, \tau \rangle$, *where $\tau = (H_1 + H_2)E_1E_2$.*

The conclusion of this section is that the effective cone $\mathrm{Eff}_3(M) = \overline{\mathrm{Eff}}_3(M)$ is generated by three extremal cycles which parametrize the following families of complete conics: H_2E_2 parametrizes reducible conics with their node lying on a fixed line, its dual H_1E_1, which parametrizes double lines with two marked-points and such that these lines pass through a fixed point, and the cycle E_1E_2 which parametrizes double lines with a unique marked-point.

4. Cycles of M of dimension 2

THEOREM 4.1. *The space $A_2(M) \otimes \mathbb{Q}$ has dimension 3. The cone of effective 2-cycles $\text{Eff}_2(M)$ is generated by the following classes*

$$\text{Eff}_2(M) = \langle H_1^2E_1, H_2^2E_2, H_1E_1E_2, H_2E_1E_2 \rangle.$$

PROOF. The space $A_2(M) \otimes \mathbb{Q}$ has dimension 3 since there are three BB-cells of dimension 2 [**6**].

It is clear that $\langle H_1^2E_1, H_2^2E_2, H_1E_1E_2, H_2E_1E_2 \rangle \subset \text{Eff}_2(M)$, and we wish to show that the opposite containment holds. In order to do that, we show the inclusion of the dual cones $\langle H_1^2E_1, H_2^2E_2, H_1E_1E_2, H_2E_1E_2 \rangle^\vee \subset \text{Eff}_2(M)^\vee$. By definition $\overline{\text{Eff}}_2(M)^\vee = \text{Nef}^2(M)$. Thus, we have the desired inclusion of dual cones as long as the classes in $\langle H_1^2E_1, H_2^2E_2, H_1E_1E_2, H_2E_1E_2 \rangle^\vee$ are nef. Nefness is what we prove next.

Note $\langle H_1^2E_1, H_2^2E_2, H_1E_1E_2, H_2E_1E_2 \rangle^\vee = \langle H_1^2, H_1H_2, H_2^2, H_1^2 - H_1H_2 + H_2^2 \rangle$. Indeed, in terms of the basis for $A_2(M) \otimes \mathbb{Q} = \langle H_1^2E_1, H_2^2E_2, H_1^3 \rangle$, we want to find $a, b, c \in \mathbb{Q}$ such that

$$H_1^2.(aH_1^2E_1 + bH_2^2E_2 + cH_1^3) = 4b + c \geq 0,$$

$$H_1H_2.(aH_1^2E_1 + bH_2^2E_2 + cH_1^3) = 2c \geq 0,$$

$$H_2^2.(aH_1^2E_1 + bH_2^2E_2 + cH_1^3) = 4a + 4c \geq 0,$$

$$(H_1^2 - H_1H_2 + H_2^2).(aH_1^2E_1 + bH_2^2E_2 + cH_1^3) = 4a + 4b + 3c \geq 0.$$

Consequently, $\langle H_1^2, H_1H_2, H_2^2, H_1^2 - H_1H_2 + H_2^2 \rangle^\vee = \langle H_1^2E_1, H_2^2E_2, H_1E_1E_2, H_2E_1E_2 \rangle$.

Since the classes H_1^2, H_1H_2, H_2^2 are nef, in order to finish the proof, it suffices to prove that $\beta = H_1^2 - H_1H_2 + H_2^2$ is also a nef class, which is Lemma 4.2. □

LEMMA 4.2. *The class $\beta = H_1^2 - H_1H_2 + H_2^2$ is a nef class in $A_3(M) \otimes \mathbb{Q}$.*

PROOF. Throughout this proof Z denotes an irreducible surface in M. We wish to show that $Z.\beta \geq 0$ for any Z.

The space M is stratified into $SL_3\mathbb{C}$-orbits and we can differentiate the irreducible surfaces in M based on how they intersect these strata. We will prove this Lemma by considering all the possibilities for these intersections and showing that in each of them, we have that $Z.\beta \geq 0$. The first case is the following. Suppose Z intersects transversally all the strata, thus the intersection $Z \cap E_1$ and $Z \cap E_2$ is transversal. Moreover, suppose $\dim Z \cap E_i = 1$. This implies that $Z.2\beta \geq 0$, as $Z.2\beta = Z.(H_1E_1 + H_2E_2) = (Z.H_1)_{|E_1} + (Z.H_2)_{|E_2}$ and $(H_i)_{|E_i}$ are nef divisors in each E_i.

The following table displays the cases to be analyzed.

case	$\dim Z \cap E_1$	$\dim Z \cap E_2$
I	1	1
II	2	2
III	2	1

(I) In case $\dim Z \cap E_1 = 1$ and $\dim Z \cap E_2 = 1$, then there exists $g \in U \subset SL_3$ such that the translate $Z^g \cap E_i$ is an irreducible curve [**9**]. At this point, we may apply the argument above and conclude that $Z^g.\beta \geq 0$, which implies that $Z.\beta \geq 0$.

(II) If dim $Z \cap E_1 = 2$ and dim $Z \cap E_2 = 2$, then Z is a divisor in E_1E_2. Hence, we can write Z as the following nonnegative linear combination $Z = aH_1E_1E_2 + bH_2E_1E_2$, for $a, b \geq 0$. Now, we compute

$$Z.\beta = (aH_1E_1E_2 + bH_2E_1E_2).(H_1^2 - H_1H_2 + H_2^2) = 0.$$

(III) There is a symmetry in the table above, hence it suffices to consider the case, dim $Z \cap E_1 = 2$ and dim $Z \cap E_2 = 1$.

Since dim $Z \cap E_1 = 2$ and Z is irreducible, then a $Z^g \subset E_1$, where Z^g is a translate. Indeed, the boundary of E_1° is the unique closed SL_3-orbit E_1E_2. Hence, the closure of $Z^g \cap E$ lies in E. Consequently, by Lemma 4.3, we can write $[Z^g]$ as the following nonnegative linear combination $Z^g = aH_1^2E_1 + bH_1E_1E_2 + cH_2E_1E_2$ for $a, b, c \geq 0$. We finish the proof by computing

$$Z^g.\beta = (aH_1^2E_1 + bH_1E_1E_2 + cH_2E_1E_2).(H_1^2 - H_1H_2 + H_2^2) = 0.$$

□

LEMMA 4.3. *The space $A_2(E_1) \otimes \mathbb{Q}$ has dimension 3. The cone of effective 2-cycles of E_1 is generated by the following classes*

$$\mathrm{Eff}_2(E_1) = \langle H_1^2E_1, H_1E_1E_2, H_2E_1E_2 \rangle.$$

PROOF. The divisor $E_1 \subset M$ is T-invariant. Following the argument as in Section 2, we observe that the Białynicki-Birula decomposition of E_1, with respect to the restriction of the T-action on E_1, has three BB-cells of dimension 2, hence the dimension of $A_2(E_1) \otimes \mathbb{Q}$ is 3.

The cycles classes $H_1^2E_1, H_1E_1E_2$ and $H_2E_1E_2$ are extremal effective cycles for the following reason. A generic point $c \in E_1$ parametrizes a double line l, with two marked-points. Thus, we have a morphism $\pi_1 : E_1 \to \mathbb{P}^{2*}$ by sending $c \mapsto l$. The class $H_1^2E_1 = (H_1^2)_{|E_1}$, in $A_2(E_1)$, is proportional to the class of the fiber $\pi^*(pt)$, which makes $H_1^2E_1$ an extremal effective cycle. On the other hand, E_1E_2 is isomorphic to the flag variety

$$\mathrm{Fl}(0, 1) = \{(p, l) | p \in l\} \subset \mathbb{P}^2 \times \mathbb{P}^{2*},$$

where the classes $H_1E_1E_2 = (H_1)_{|E_1E_2}$ and $H_2E_1E_2 = (H_2)_{|E_1E_2}$, in $A_2(E_1E_2)$, are proportional to the classes of the fibers of the projection morphisms into $\mathbb{P}^2$ and $\mathbb{P}^{2*}$. This implies that they are extremal effective cycles in $A_2(E_1E_2)$. Now, the inclusion morphism $\iota : E_1E_2 \to M$ induces an injective map $\iota_* : A_2(E_1E_2) \to A_2(M)$. By definition of M, there are two projection maps $\pi_i : M \to \mathbb{P}^5$, $(i = 1, 2)$, and observe that the classes $H_1E_1E_2$ and $H_2E_1E_2$ are contracted by them. It follows that $H_1E_1E_2$ and $H_2E_1E_2$ are extremal cycles in $A_2(E_1)$ [**3**]. In order to show that these three cycles are all the extremal cycles, we compute the dual cone $\langle H_1^2E_1, H_1E_1E_2, H_2E_1E_2 \rangle^\vee$ and show it is generated by nef cycle classes. Observe that Lemma 3.4 carries this out partly (it shows that the wall generated by $H_1E_1E_2$ and $H_2E_1E_2$ is extremal), following the same strategy as in Lemma 3.2. We omit the details as no difficulty arises. □

COROLLARY 4.4. *The cone of numerically effective* 3*-cycles is generated by the following classes* $\mathrm{Nef}^2(M) = \langle H_1^2, H_2^2, H_1H_2, \beta \rangle$, *where* $\beta = H_1^2 - H_1H_2 + H_2^2$.

The class $H_1^2E_1$ parametrizes fixed double lines with two marked-points which vary. Hence, $H_1^2E_1 \cong \mathbb{P}^{1[2]}$. The class $H_2E_1E_2$ parametrizes double lines such that the unique marked-point in it lies in a fixed line. Hence, $H_2E_1E_2 \cong \mathbb{P}^{2*}$.

It is important to mention that Theorem 3.1 and Theorem 4.1 are consistent with the theory of spherical varieties. For example, both of the subvarieties in the previous paragraph are rational as the theory of spherical varieties predicted [**12**].

Acknowledgments

I thank Izzet Coskun for suggesting the proof of the main result of this note. Thanks to Dawei Chen, Joe Harris and Francesco Cavazzani for useful conversations. I would also like to thank Chris Gomes for reading a draft of this note and generously improving the language of it. I thank also to the referee for a detailed reading of this note and useful suggestions.

References

[1] A. Białynicki-Birula, *Some theorems on actions of algebraic groups*, Ann. of Math. (2) **98** (1973), 480–497. MR0366940

[2] M. Chasles, *Determination du nombre de sections coniques qui doivent toucher cinq courbes données d'ordre quelconque, ou satisfaire à diverses autres conditions*, C.R. Acad. Sci. Paris **58** (1864), pp. 222–226.

[3] Dawei Chen and Izzet Coskun, *Extremal higher codimension cycles on moduli spaces of curves*, Proc. Lond. Math. Soc. (3) **111** (2015), no. 1, 181–204. MR3404780

[4] Olivier Debarre, Lawrence Ein, Robert Lazarsfeld, and Claire Voisin, *Pseudoeffective and nef classes on abelian varieties*, Compos. Math. **147** (2011), no. 6, 1793–1818. MR2862063

[5] C. De Concini and C. Procesi, *Complete symmetric varieties*, Invariant theory (Montecatini, 1982), Lecture Notes in Math., vol. 996, Springer, Berlin, 1983, pp. 1–44. MR718125

[6] William Fulton, *Intersection theory*, 2nd ed., Ergebnisse der Mathematik und ihrer Grenzgebiete. 3. Folge. A Series of Modern Surveys in Mathematics [Results in Mathematics and Related Areas. 3rd Series. A Series of Modern Surveys in Mathematics], vol. 2, Springer-Verlag, Berlin, 1998. MR1644323

[7] Phillip Griffiths and Joseph Harris, *Principles of algebraic geometry*, Wiley-Interscience [John Wiley & Sons], New York, 1978. Pure and Applied Mathematics. MR507725

[8] Joe Harris, *Algebraic geometry*, Graduate Texts in Mathematics, vol. 133, Springer-Verlag, New York, 1992. A first course. MR1182558

[9] Steven L. Kleiman, *The transversality of a general translate*, Compositio Math. **28** (1974), 287–297. MR0360616

[10] Steven L. Kleiman, *Intersection theory and enumerative geometry: a decade in review*, Algebraic geometry, Bowdoin, 1985 (Brunswick, Maine, 1985), Proc. Sympos. Pure Math., vol. 46, Amer. Math. Soc., Providence, RI, 1987, pp. 321–370. With the collaboration of Anders Thorup on §3. MR927987

[11] D. Mumford, *Rational equivalence of* 0*-cycles on surfaces*, J. Math. Kyoto Univ. **9** (1968), 195–204. MR0249428

[12] Nicolas Perrin, *On the geometry of spherical varieties*, Transform. Groups **19** (2014), no. 1, 171–223. MR3177371

[13] Francesco Severi, *Serie, Sistemi d'Equivalenza e Corrispondenze Algebriche sulle Varietà Algebriche. Vol. 1* (Italian), Edizioni Cremonense, Rome, 1942. MR0026819

DEPARTMENT OF MATHEMATICS, HARVARD UNIVERSITY, CAMBRIDGE, MASSACHUSETTS 02138
Current address: Instituto de Matemáticas, Universidad Nacional Autónoma de México, Oaxaca de Juárez, Oax, México
Email address: lozano@im.unam.mx

Contemporary Mathematics
Volume **709**, 2018
http://dx.doi.org/10.1090/conm/709/14295

On Shimura's isomorphism and (Γ, G)-bundles on the upper-half plane

Claudio Meneses

ABSTRACT. For a compact real form U of a complex simple group G, and an irreducible representation $\rho : \Gamma \to U$ of a Fuchsian group of the first kind Γ, it is shown that the classical isomorphism of Shimura, for the periods of a cusp form of weight 2 with values in $\mathfrak{g}$ and the representation $\mathrm{Ad}\rho : \Gamma \to \mathrm{Aut}\mathfrak{g}$, can be interpreted as the differential at a point of the zero section, for a natural map from the cotangent bundle of the moduli space of certain (Γ, G)-bundles over $\mathbb{H}$ (in the sense of Seshadri) to an open set in the smooth locus of the character variety $\mathrm{Hom}_{\mathfrak{t}}(\Gamma, G)/PG$. Emphasis is put on analytic techniques.

CONTENTS

1. Introduction

Let G be a complex finite-dimensional simple group with Lie algebra $\mathfrak{g}$, $U \subset G$ a compact real form, and $\rho : \Gamma \to U$ an irreducible representation of an arbitrary Fuchsian group of the first kind Γ. A *$\mathfrak{g}$-valued cusp form of weight 2 with the representation* $\mathrm{Ad}\,\rho$ is a holomorphic function $\Phi : \mathbb{H} \to \mathfrak{g}$ satisfying the functional equation

$$\Phi(\gamma\tau)\gamma'(\tau) = \mathrm{Ad}\,\rho(\gamma) \cdot \Phi(\tau), \qquad \forall\gamma \in \Gamma,\ \tau \in \mathbb{H}, \tag{1.1}$$

where $\mathrm{Ad}\,\rho$ is the composition of ρ and the adjoint representation of G in $\mathfrak{g}$, such that for every cusp $\tau_i \in \mathbb{R} \cup \{\infty\}$,

$$\lim_{\tau \to \tau_i} \Phi(\tau) = 0, \tag{1.2}$$

where the limit is understood to be taken within a fundamental region $\mathfrak{F}$. The spaces $\mathfrak{S}_2(\Gamma, \mathrm{Ad}\,\rho)$ of cusp forms of weight 2 (and of arbitrary positive even weight) were introduced by Shimura [**Shi71**], are finite dimensional, and thanks to the Killing

2010 *Mathematics Subject Classification.* Primary 30F35, 32G13; Secondary 20J06.

form, admit a Hermitian inner product generalizing the Petersson inner product [**Pet40, Pet48**].

The author's interest in the subject stems from the humble realization that the theory of moduli of vector bundles with parabolic structures over Riemann surfaces, as in the work of Narasimhan-Mehta-Seshadri [**NS65, MS80**], can be fully understood in terms of complex analysis on the upper half-plane, and more concretely, the harmonic theory for families of Cauchy-Riemann operators. Such a complex-analytic approach has been pursued, for instance, in [**TZ08**], and is sufficiently flexible to allow the replacement of the groups $\mathrm{U}(r) \subset \mathrm{GL}(r, \mathbb{C})$, by general simple pairs $U \subset G$, and their corresponding geometric structures. These are the objects we will concentrate our attention on in this paper.

Ideally, one wishes to associate, to every irreducible $\rho : \Gamma \to U$, a holomorphic principal G-bundle over the compact Riemann surface $\Gamma \setminus \mathbb{H}^*$, with an additional parabolic structure on each cusp and elliptic fixed point, a reduction of the structure group to U away from the cusps and elliptic fixed points, and a singular flat U-connection whose holonomy correspond to the conjugacy class of ρ. One would then expect a correspondence between the set of such ρ and the set of stable parabolic G-bundles. While this is true for stable principal G-bundles [**Ram75**], it turns out that such association is not possible in general for Fuchsian groups whose elements are not purely hyperbolic, and only in the generic case when the weights (for all parabolic and elliptic generators) lie in the interior of the Weyl alcove. In general, the geometric structures suitable for this purpose have been called *parahoric G-torsors* by Balaji-Seshadri (see [**BS15**] and references therein), and indeed satisfy the previously mentioned uniformization principle, generalizing the results in [**NS65, MS80**]. Our perspective is based on working over $\mathbb{H}$ and directly considering, instead, some induced spaces of global sections (in the sense of sheaf cohomology), as suitable spaces of automorphic forms.

In a seemingly unrelated line, the study of arithmetic properties of automorphic functions let Goro Shimura to generalize the work of Eichler to periods associated to representations of Fuchsian groups on compact groups. For weight 2, the celebrated isomorphism of Shimura [**Shi71**] reads

$$\mathfrak{S}_2(\Gamma, \mathrm{Ad}\,\rho) \oplus \overline{\mathfrak{S}_2(\Gamma, \mathrm{Ad}\,\rho)} \cong H^1_P(\Gamma, \mathfrak{g}_{\mathrm{Ad}\,\rho}).$$

One is undoubtedly left to notice that, on one hand, the parabolic cohomology $H^1_P(\Gamma, \mathfrak{g}_{\mathrm{Ad}\,\rho})$ models the tangent space at an equivalence class $\{\rho\}$ of representations in a complex character variety $\mathscr{K}_{\mathbb{C}}$, and on the the other, the spaces $\mathfrak{S}_2(\Gamma, \mathrm{Ad}\,\rho)$ and $\overline{\mathfrak{S}_2(\Gamma, \mathrm{Ad}\,\rho)}$ model, respectively, the tangent and cotangent spaces at a point in a moduli space $\mathscr{N}$ of (Γ, G)-principal bundles over $\mathbb{H}$ [**TZ08**] (cf. [**Iva88**]).

The previous picture is completed after one notices it is possible to associate, in a natural and injective way, an irreducible complex representation $\chi : \Gamma \to G$ to a pair consisting of an irreducible representation $\rho : \Gamma \to U$, and an element $\Phi \in \mathfrak{S}_2(\Gamma, \mathrm{Ad}\,\rho)$ (see section 4), in such a way that the (Γ, G)-bundles over $\mathbb{H}$ induced by χ and ρ are equivalent.[1] We refer to such correspondence as the *Narasimhan-Mehta-Seshadri map*. Considering the zero section s_0 of the cotangent bundle $T^*\mathscr{N}$, induces a canonical splitting

$$s_0^*\,(T(T^*\mathscr{N})) \cong T^*\mathscr{N} \oplus T\mathscr{N},$$

[1] A word of caution: χ *is not the same* than the induced representation considered in the study of harmonic metrics when $G = \mathrm{GL}(r, \mathbb{C})$.

as vector bundles over $\mathscr{N}$. Thus, Shimura's work on parabolic cohomology [**Shi71**] admits a natural geometric interpretation in the previous setup (a fact observed by Goldman [**Gol84**] for the group $\mathrm{PSL}(2,\mathbb{R})$, scalar cusp forms of higher weight, and their Eichler periods). The main result of this work is the following theorem.

THEOREM 1. *With respect to the complex coordinates given by the choice of an arbitrary basis in $\mathfrak{S}_2(\Gamma, \operatorname{Ad}\rho) \oplus \overline{\mathfrak{S}_2(\Gamma, \operatorname{Ad}\rho)}$, the differential of the Narasimhan-Mehta-Seshadri map $\mathscr{F} : T^*\mathscr{N}^s \to \mathscr{K}_{\mathbb{C}}$ at the point $\{P_\rho, 0\} \in T^*\mathscr{N}^s$ over the image of the zero section in $T^*\mathscr{N}^s$ is given as*

$$d\mathscr{F}_{\{P_\rho,0\}}(\Psi, \nu) = \mathscr{L}_S\,(\Psi, \nu)\,, \tag{1.3}$$

where $\mathscr{L}_S$ is Shimura's isomorphism.

In other words, theorem 1 states that Shimura's isomorphism [**Shi71**] between complex parabolic cohomology and spaces of $\mathfrak{g}$-valued cusp forms of weight 2 with the representation $\operatorname{Ad}\rho$ is essentially the infinitesimal version of the Narasimhan-Mehta-Seshadri map $\mathscr{F} : T^*\mathscr{N}^s \to \mathscr{K}_{\mathbb{C}}$ over s_0, a fact that illustrates explicitly the trascendental nature of $\mathscr{F}$. In fact, it is possible to say more: $\mathscr{F}$ is a real-analytic monomorphism, and $\mathscr{F} \circ s_0$ is also a symplectomorphism onto its image, once the unitary character variety $\mathscr{K}^s$ is endowed with Goldman's symplectic form, and the natural Kähler structure induced from Petersson inner product is considered over $\mathscr{N}^s$.

The contents of this work are organized as follows: section 2 presents the conventions in which the present work is founded; section 3 describes the construction of the character variety $\mathrm{Hom}_{\mathfrak{t}}(\Gamma, G)/G$; in section 4, we develop the infinitesimal deformation theory of (Γ, G)-bundles from a complex analytic perspective. Finally, a proof of theorem 1 (and some corollaries) is presented in section 5.

I am deeply grateful to the Max Planck Institute for Mathematics in Bonn, where the present work was developed, for its generosity and stimulating working atmosphere, and to CIMAT in Mexico, where the manuscript was put in final form. I would also like to warmly thank Indranil Biswas for pointing out an important issue in the first version of this manuscript, and the referee for the useful comments provided.

2. Lie algebra-valued cusp forms of weight 2

Let Γ be a Fuchsian group (i.e. a discrete subgroup of $\mathrm{PSL}(2,\mathbb{R})$) of the first kind. It is a classical result [**Hej83, Leh64**] that Γ is finitely generated and admits an explicit presentation in terms of $2g$ hyperbolic generators $A_1, \dots, A_g, B_1, \dots, B_g$, m elliptic generators $S_1, \dots, S_m$, and n parabolic generators $T_{m+1}, \cdots, T_{m+n}$, with relations

$$\prod_{i=1}^{g}[A_i, B_i] \cdot \prod_{j=1}^{m} S_j \cdot \prod_{k=1}^{n} T_{m+k} = e,$$

$$S_1^{c_1} = \cdots = S_m^{c_m} = e,$$

for some integers $2 \le c_j < \infty$. When talking about a Fuchsian group, we will assume to have fixed once and for all one such *marking*.

The *signature* of Γ is the collection of nonnegative integers $(g; c_1, \ldots, c_m; n)$, $j = 1, \ldots, m$, subject to the hyperbolicity condition

$$2g - 2 + \sum_{i=1}^{m} \left(1 - \frac{1}{c_i}\right) + n > 0.$$

The equivalence relation on Fuchsian groups is conjugation in $\mathrm{PSL}(2,\mathbb{R})$. Γ admits a fundamental region $\mathfrak{F}$ of finite hyperbolic area, bounded by a finite number of geodesic arcs and vertices determined by the fixed points of the generators $\{S_1, \ldots, S_m, T_{m+1}, \ldots, T_{m+n}\}$. We will denote by $\{e_1, \cdots, e_m\} \subset \mathbb{H}$ the set of elliptic fixed points ($S_i(e_i) = e_i$), and by $\{p_1, \cdots, p_n\} \subset \mathbb{R} \cup \{\infty\}$ the set of cusps ($T_{m+i}(p_i) = p_i$) in the fundamental region $\mathfrak{F}$. We follow Shimura's convention [**Shi71**] on the construction of the compact quotient

$$S = \Gamma \setminus \mathbb{H}^*$$

with local elliptic (resp. parabolic) coordinates $\zeta_i = \zeta^{c_i} \circ \sigma_i^{-1}$ (resp. $q_i = q \circ \sigma_i^{-1}$) where $q = e^{2\pi\sqrt{-1}\tau}$, $\zeta : \mathbb{H} \to \mathbb{D}$ is the Cayley mapping, and $\sigma_1, \ldots, \sigma_{m+n} \in \mathrm{PSL}(2,\mathbb{R})$ satisfy

$$\zeta\left((\sigma_i^{-1} S_i \sigma_i)(\tau)\right) = e^{2\pi\sqrt{-1}/c_i} \cdot \zeta(\tau), \qquad i \le m \tag{2.1}$$

$$\left(\sigma_i^{-1} T_i \sigma_i\right)(\tau) = \tau \pm 1, \qquad i > m. \tag{2.2}$$

2.1. Spaces of $\mathfrak{g}$-valued cusp forms. Let $\rho : \Gamma \to U$ be irreducible with induced complex irreducible representation $\mathrm{Ad}\,\rho : \Gamma \to \mathrm{Aut}\,\mathfrak{g}$, $p, q \in \{0, 1\}$, and let us denote by $\mathscr{C}_c^{p,q}(\Gamma, \mathrm{Ad}\,\rho)$ the space of smooth, $\mathfrak{g}$-valued functions on $\mathbb{H}$, compactly supported within a fundamental region $\mathfrak{F}$ for Γ, that satisfy

$$\Phi(\gamma\tau)\gamma'(\tau)^p \overline{\gamma'(\tau)^q} = \mathrm{Ad}\,\rho(\gamma) \cdot \Phi(\tau) \qquad \forall\, \tau \in \mathbb{H},\ \gamma \in \Gamma,$$

For any $\Phi_1, \Phi_2 \in \mathscr{C}_c^{p,q}(\Gamma, \mathrm{Ad}\,\rho)$, their *Petersson inner product* is defined as

$$\langle \Phi_1, \Phi_2 \rangle_P = \iint_{\mathfrak{F}} \langle \Phi_1, \Phi_2 \rangle \, y^{2p+2q-2} dx dy \tag{2.3}$$

where $\tau = x + \sqrt{-1}y$, and $\langle \Phi_1, \Phi_2 \rangle = -K\left(\Phi_1, \overline{\Phi_2}\right)$ is, pointwise, the Hermitian inner product in $\mathfrak{g}$ determined by the Killing form K and the Cartan involution $X \mapsto \overline{X}$ defined by complex conjugation of the realization $\mathfrak{g} \cong \mathfrak{u}_{\mathbb{C}}$ (for $\mathfrak{u}$ the Lie algebra of U) [**Hel79**]. A fundamental property of the Petersson inner product is its invariance under the induced action of Γ in $\mathscr{C}_c^{p,q}(\Gamma, \mathrm{Ad}\,\rho)$. Let us denote by $\mathscr{H}^{p,q}(\Gamma, \mathrm{Ad}\,\rho)$ the completion of $\mathscr{C}_c^{p,q}(\Gamma, \mathrm{Ad}\,\rho)$ to a Hilbert space under the Petersson inner product. In terms of the Cauchy-Riemann operators

$$\bar{\partial}^{p,q} : \mathscr{H}^{p,q}(\Gamma, \mathrm{Ad}\,\rho) \to \mathscr{H}^{p,q+1}(\Gamma, \mathrm{Ad}\,\rho),$$

it is also possible to define the spaces of $\mathfrak{g}$-valued cusp forms of weight 2 as

$$\mathfrak{S}_2(\Gamma, \mathrm{Ad}\,\rho) = \ker \bar{\partial}^{1,0}.$$

3. Character varieties for simple groups

We recall the construction of the smooth locus of a character variety for a Fuchsian group Γ and a simple group G. For details, the reader is referred to [**Gol84**] and references therein.

Let $T \subset U$ be a choice of maximal torus for $U \subset G$, and let $\mathfrak{t} \subset \mathfrak{u}$ denote its Lie algebra. Let us fix, for each parabolic and elliptic generator of Γ, an element $t_i \in \mathfrak{t}$

and its induced adjoint orbit in G, $\mathcal{O}_i = [\exp t_i] \subset G$, $i = 1, \ldots, m+n$. We will denote $\mathbf{t} = \{t_i\}_{i=1}^{m+n}$, a collection corresponding to a choice of weights. Consider the subset $\mathrm{Hom}_{\mathbf{t}}(\Gamma, G) \subset \mathrm{Hom}(\Gamma, G)$ consisting of representations $\chi : \Gamma \to G$ with $\chi(S_i) \in \mathcal{O}_i$, $\chi(T_i) \in \mathcal{O}_i$. The smooth locus $\mathscr{P}$ in $\mathrm{Hom}_{\mathbf{t}}(\Gamma, G)$ is the collection of representations $\chi : \Gamma \to G$ satisfying

$$\dim H^0(\Gamma, \mathrm{Ad}\,\chi) = \dim \mathfrak{g}^{\mathrm{Ad}\,\chi} = \dim \mathfrak{g}^{\mathrm{Ad}\,G} = 0,$$

which is an open condition. More specifically, $\mathrm{Hom}_{\mathbf{t}}(\Gamma, G) = \psi^{-1}(I)$, where ψ is the map

$$\psi : G^{2g} \times \prod_{i=1}^{n+m} \mathcal{O}^i \to U,$$

$$(M_1, N_1, \ldots, M_g, N_g, f_1, \ldots, f_{n+m}) \mapsto \prod_{i=1}^{g} [M_i, N_i] \prod_{i=1}^{n+m} f_i,$$

and the smooth locus is the subset $\mathscr{P} \subset \psi^{-1}(I)$ of regular points of ψ, consisting of representations χ for which $\dim Z(\chi) = \dim Z(G) = 0$. Even though the action of $PG = G/Z(G)$ on $\mathscr{P}$ by conjugation is locally free and generically free, it is not proper in general. The Zariski open set $\mathrm{Hom}^0_{\mathbf{t}}(\Gamma, G)$ in $\mathscr{P}$ where G acts freely and properly is the space of χ whose image in G does not lie in a group of the form $P \times Z(G)$, for P a parabolic subgroup of $[G, G] = G$. Such a restriction leads to the construction of the *smooth locus*

$$\mathscr{K}_{\mathbb{C}} = \mathrm{Hom}^0_{\mathbf{t}}(\Gamma, G)/PG$$

of the character variety $\mathrm{Hom}_{\mathbf{t}}(\Gamma, G)/PG$. Since G is simple, if nonempty, $\mathscr{K}_{\mathbb{C}}$ is a complex-analytic manifold of dimension

$$2(g-1)\dim_{\mathbb{C}} G + \sum_{i=1}^{n+m} \dim_{\mathbb{C}} \mathcal{O}^i,$$

where $\dim_{\mathbb{C}} \mathcal{O}_i = \mathrm{rank}(\mathrm{Ad}\exp(t_i) - I)$ on $\mathfrak{g}$. The infinitesimal deformations of a given $\chi \in \mathrm{Hom}^0_{\mathbf{t}}(\Gamma, G)$, leaving fixed the conjugacy classes of its elliptic and parabolic generators, are determined by the space of *parabolic 1-cocycles* $Z^1_P(\Gamma, \mathfrak{g}_{\mathrm{Ad}\,\chi})$, consisting of maps $z : \Gamma \to \mathfrak{g}$ satisfying

$$z(\gamma_1\gamma_2) = z(\gamma_1) + \mathrm{Ad}\,\chi(\gamma_1) \cdot z(\gamma_2), \quad \forall\, \gamma_1, \gamma_2 \in \Gamma$$

and

$$z(S_i) \in (\mathrm{Ad}\,\chi(S_i) - I)\mathfrak{g}, \qquad z(T_j) \in (\mathrm{Ad}\,\chi(T_j) - I)\mathfrak{g}.$$

The infinitesimal deformations tangential to the PG-adjoint orbit of χ in $\mathrm{Hom}^0_{\mathbf{t}}(\Gamma, G)$ are precisely the 1-coboundary maps $z(\gamma) = (\mathrm{Ad}\,\chi(\gamma) - I)X$, $X \in \mathfrak{g}$, and hence the tangent space at the equivalence class $\{\chi\} \in \mathscr{K}_{\mathbb{C}}$ is modeled by the *1st parabolic cohomology space* $H^1_P(\Gamma, \mathfrak{g}_{\mathrm{Ad}\,\chi})$ (see [**Shi71**] for details on these spaces; their dimensions were calculated by Weil [**Wei64**]).

REMARK 1. In general, a representation $\chi \in \mathrm{Hom}^0_{\mathbf{t}}(\Gamma, G)$ need not be irreducible. The collection of equivalence classes of irreducible representations (without inner automorphisms in PG) forms an open set in $\mathscr{K}_{\mathbb{C}}$.

Repeating the construction with U instead, we conclude that there is a real-analytic manifold $\mathscr{K}^s$ of real dimension $\dim_{\mathbb{C}} \mathscr{K}_{\mathbb{C}}$, consisting of orbits of irreducible representations $\rho : \Gamma \to U$ with prescribed conjugacy classes of parabolic and elliptic generators (where PU acts freely and without adjoint invariants in $\mathfrak{u}$), since it readily follows that an irreducible representation $\rho : \Gamma \to U$ cannot have inner automorphisms in PU. In such case, the possible values of parabolic and elliptic elements are parametrized by the generalized flag manifolds $\mathscr{F}^i = U/Z\,(\exp t_i)$ associated with each weight $t_i \in \mathfrak{t}$, and there is an inclusion $\mathscr{K}^s \hookrightarrow \mathscr{K}_{\mathbb{C}}$. In turn, the tangent space at an equivalence class $\{\rho\} \in \mathscr{K}^s$ is modeled by the space $H^1_P(\Gamma, \mathfrak{u}_{\mathrm{Ad}\,\rho})$.

4. Deformation theory of (Γ, G)-principal bundles on $\mathbb{H}$

Given a representation $\chi : \Gamma \to G$, *the induced* (Γ, G)*-bundle on* $\mathbb{H}$ is the trivial bundle $\mathbb{H} \times G$ with χ acting as bundle automorphisms commuting with the right G-action [**BS15, Ses10**]. Hence, two (Γ, G)-bundles induced by χ_1, χ_2 are *equivalent* if there exists a holomorphic map $f : \mathbb{H} \to G$, satisfying $f(\gamma\tau) = \chi_1(\gamma) f(\tau) \chi_2(\gamma)^{-1}$ $\forall \gamma \in \Gamma$, $\tau \in \mathbb{H}$, and whose limit exists at every cusp.[2] In particular, the harmonic theory implies that a (Γ, G)-bundle induced by χ is isomorphic to at most one induced by a unitary representation $\rho : \Gamma \to U$, up to conjugation in U. Let $\{P_\rho\}$ denote the class of (Γ, G)-bundles on $\mathbb{H}$ induced by a class of irreducible representations $[\rho] \in \mathscr{K}^s$ (that is, we assume that no representative ρ has inner automorphisms in PU). Our aim is to introduce complex coordinates and a manifold structure over the set $\mathscr{N}^s = \{\{P_\rho\}_{[\rho] \in \mathscr{K}}\}$. The results of the present section are a generalization of the analysis developed in [**TZ08**] for stable parabolic bundles to (Γ, G)-bundles on $\mathbb{H}$, and are, in turn, inspired by the theory of quasiconformal mappings of Ahlfors-Bers. Thanks to [**BS15**], the resulting complex manifold corresponds to the moduli space of stable parahoric G-torsors with prescribed parabolic structure whose automorphism group is equal to $Z(G)$.

On the analytic side, by considering open sets in $\mathbb{H}$ invariant under Γ, one can define the sheaf $\mathbf{P}_\rho$ on $\mathbb{H}$, of holomorphic G-valued maps which are automorphic with respect to ρ, and are bounded at the cusps. By the general deformation theory of Kodaira-Spencer, its infinitesimal deformations are encoded in the cohomology

$$H^1(\mathbb{H}, \mathbf{E}_{\mathrm{Ad}\,\rho}),$$

where $\mathbf{E}_{\mathrm{Ad}\,\rho}$ is the sheaf of holomorphic $\mathfrak{g}$-valued maps on $\mathbb{H}$ which are automorphic with respect to the composition $\mathrm{Ad}\,\rho = \mathrm{Ad} \circ \rho$ and bounded at the cusps. Such cohomology is naturally isomorphic, by Serre duality, to the dual space $\mathfrak{S}_2(\Gamma, \mathrm{Ad}\,\rho)^\vee$, and models the holomorphic cotangent space at $\{P_\rho\}$ (cf. [**Men17**], where the proof for the unitary case is sketched in detail; the argument for a general simple group G is essentially the same, at least when the parabolic and elliptic weights are generic, and there is an underlying principal bundle with parabolic structure over $S = \Gamma \setminus \mathbb{H}^*$).[3]

We have pointed out that the space $\mathfrak{S}_2(\Gamma, \mathrm{Ad}\,\rho)$ of cusp forms of weight 2 corresponds to the kernel of the Cauchy-Riemann operator over the Hilbert space $\mathscr{H}^{1,0}(\Gamma, \mathrm{Ad}\,\rho)$. After consideration of duality and the Petersson inner product,

[2]Observe that, in particular, the conjugacy classes of χ_1 and χ_2 at every parabolic and elliptic generator are necessarily the same.

[3]The geometric interpretation of $\mathfrak{g}$-valued cusp forms of weight 2 as global holomorphic sections of the sheaf $\mathbf{E}^\vee_{\mathrm{Ad}\,\rho} \otimes K_S$ enables the computation of the dimension of the space $\mathfrak{S}_2(\Gamma, \mathrm{Ad}\,\rho)$ by means of the Riemann-Roch theorem (cf. [**Men17**]).

such fact is consistent with the Dolbeaut isomorphism of sheaves

$$H^1(\mathbb{H}, \mathbf{E}_{\operatorname{Ad}\rho}) \cong H^{0,1}_{\bar\partial}(\mathbb{H}, \mathbf{E}_{\operatorname{Ad}\rho}).$$

Moreover, thanks to the Petersson inner product, the previous isomorphism can be enhanced if one considers the Hodge Laplacian $\Delta = \bar\partial^*$ over $\mathscr{H}^{1,0}(\Gamma, \operatorname{Ad}\rho)$; there is an additional isomorphism of the previous spaces with the space of harmonic $(0,1)$-forms with values in $\mathfrak{g}$, which are $\operatorname{Ad}\rho$-invariant and vanish at the cusps. Now, since the family of (Γ, G)-bundles on $\mathbb{H}$ that we are considering is parametrized by the space $\mathscr{K}^s$, for every ρ, there is an induced Kodaira-Spencer map

$$H^1_P(\Gamma, \mathfrak{u}_{\operatorname{Ad}\rho}) \to H^1(\mathbb{H}, \mathbf{E}_{\operatorname{Ad}\rho}),$$

and the former space is, by the classical theorem of Shimura, isomorphic to the space $\mathfrak{S}_2(\Gamma, \operatorname{Ad}\rho)$. Note that, as stated, Shimura's isomorphism is a real isomorphism.

The following diagram summarizes and illustrates the role that the cusp forms of weight 2 and Shimura's isomorphism play in this dual set-up:

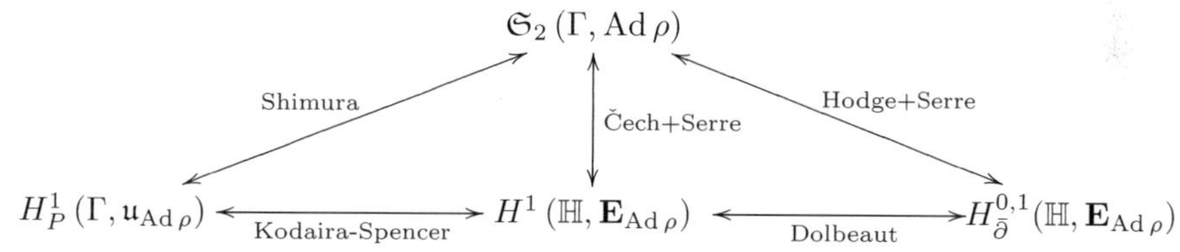

Let $\mathscr{N}^s$ be the set of equivalence classes of (Γ, G)-bundles on $\mathbb{H}$ of the form P_ρ, for $[\rho] \in \mathscr{K}^s$. We wish to turn $\mathscr{N}^s$ into a complex manifold, in such a way that the holomorphic tangent space at a point $\{P_\rho\} \in \mathscr{N}^s$ is modeled by the spaces $\overline{\mathfrak{S}_2(\Gamma, \operatorname{Ad}\rho)} \cong \mathfrak{S}_2(\Gamma, \operatorname{Ad}\rho)^\vee$, for any representative ρ (observe that such spaces are isomorphic under conjugation). In order to introduce complex coordinates in $\mathscr{N}^s$, we need to provide a method of construction of local deformations of Cauchy-Riemann operators. The fundamental result to provide a complex-analytic description of the deformation theory of (Γ, G)-bundles is the following (cf. [**MN84**]).

LEMMA 1. *Let $\rho : \Gamma \to U$ be an irreducible representation in $\mathscr{K}^s$. For every $\nu \in \mathfrak{S}_2(\Gamma, \operatorname{Ad}\rho)$ and $\epsilon \in \mathbb{C}$ sufficiently close to 0, there is a function $f^{\epsilon\nu} : \mathbb{H} \to G$ solving the differential equation $f^{-1} f_{\bar\tau} = \epsilon\nu$, satisfying*

$$f^{\epsilon\nu}(\gamma\tau) = \rho^{\epsilon\nu}(\gamma) f^{\epsilon\nu}(\tau) \rho(\gamma)^{-1} \tag{4.1}$$

for some irreducible representation $\rho^{\epsilon\nu} : \Gamma \to U$ in $\mathscr{K}^s$, real-analytic and regular at each cusp.The set $\{f^{\epsilon\nu}\}$ of such solutions is a torsor for the left action of U. moreover, the conjugacy classes of parabolic and elliptic elements are preserved.

PROOF. This is essentially the same argument as in [**TZ08**], p.124: existence of antiholomorphic solutions $f : \mathbb{H} \to G$, regular at the cusps, is standard (see, for instance, [**MN84**]). Then, the equation $f^{-1} f_{\bar\tau} = \epsilon\nu$ admits an antiholomorphic solution f_-, depending holomorphically in ϵ. Consider any such f_-, and for any

$\gamma \in \Gamma$, consider the function $\chi_\gamma(\overline{\tau}) = \left(f_-^{-1} \circ \gamma\right) \rho(\gamma) f_-$. Then, since ν is $\operatorname{Ad}\rho$-automorphic,

$$\frac{\partial}{\partial \overline{\tau}} \chi_\gamma(\overline{\tau}) = \epsilon \left(f_-^{-1} \circ \gamma\right) \rho(\gamma) \nu f_- - \epsilon \left(f_-^{-1} \circ \gamma\right) (\nu \circ \gamma) \overline{\gamma'} \rho(\gamma) f_- \equiv 0.$$

Since χ_γ is antiholomorphic, it must be equal to a constant $\chi(\gamma)$, and clearly $\chi(\gamma_1\gamma_2) = \chi(\gamma_1)\chi(\gamma_2)$. Thus, the function f_- satisfies the functional equation $f_- \circ \gamma = \chi(\gamma) f_- \rho(\gamma)^{-1}$ for some representation $\chi : \Gamma \to G$, defined up to conjugation. Following the main result in [**BS15**], and by the openness of stability, for sufficiently small ϵ, the monodromy representation associated to f_- defines a (Γ, G)-bundle on $\mathbb{H}$ which is equivalent to an irreducible unitary one, and consequently, there exists a holomorphic map $f_+ : \mathbb{H} \to G$, regular at each cusp, such that $f_+ f_- = f^{\epsilon\nu} : \mathbb{H} \to G$ is a solution of (4.1), depending real-analytically in τ and ϵ, and satisfying $f^{\epsilon\nu}(\gamma\tau) = \rho^{\epsilon\nu}(\gamma) f^{\epsilon\nu}(\tau) \rho(\gamma)^{-1}$ for all $\gamma \in \Gamma$, with $\rho^{\epsilon\nu} : \Gamma \to U$ irreducible. Clearly, such $f^{\epsilon\nu}$ is defined up to left multiplication by an element in U.

Finally, the regularity of $f^{\epsilon\nu}$ at the cusps and elliptic fixed points implies that

$$\chi(S_i) = f^{\epsilon\nu}(e_i)^{-1} \rho(S_i) f^{\epsilon\nu}(e_i), \quad \chi(T_j) = f^{\epsilon\nu}(p_j)^{-1} \rho(T_j) f^{\epsilon\nu}(p_j)$$

Therefore, the conjugacy classes of such group elements are preserved under deformations. This concludes the proof. □

REMARK 2. Observe that the collection of solutions $\{f^{\epsilon\nu}\}$ in lemma 1 is a torsor for PU, and is thus an invariant of the classes $\{\rho\}$, $\{\rho^{\epsilon\nu}\}$. Such solutions define a map of (Γ, G)-bundles, and by pull-back, allow to express the Cauchy-Riemann operator of $P_{\rho^{\epsilon\nu}}$ over P_ρ as

$$(f^{\epsilon\nu})^{-1} \circ \bar{\partial} \circ f^{\epsilon\nu} = \bar{\partial} + \epsilon\nu.$$

This is our set of complex coordinates in $\mathscr{N}^s$, analogous to Bers' coordinates on Teichmüller spaces. It can be verified [**TZ08**] that such coordinates transform holomorphically, and thus turn $\mathscr{N}^s$ into a complex manifold. Its complex dimension, which depends on the fixed set of weights $\mathbf{t} = \{t_i\}_{i=1}^{m+n}$, is equal to $(g-1)\dim U + \sum_{i=1}^{m+n} \dim_{\mathbb{C}} \mathscr{F}^i$.

COROLLARY 1. *Complex coordinates, analogous to the Bers' coordinates on Teichmüller spaces, can be introduced in $\mathscr{N}^s$. Given $f^{\epsilon\nu}$ as in lemma 1, the holomorphic structure corresponding to $\rho^{\epsilon\nu}$ is realized in the Γ-vector bundle over $\mathbb{H}$ associated to the sheaf $\mathbf{E}_{\operatorname{Ad}\rho}$, by means of the $\bar{\partial}$-operator $(\operatorname{Ad} f^{\epsilon\nu})^{-1} \circ \bar{\partial} \circ \operatorname{Ad} f^{\epsilon\nu}$.*

LEMMA 2. *Let $\nu \in \overline{\mathfrak{S}_2(\Gamma, \operatorname{Ad}\rho)}$ and $\epsilon \in \mathbb{C}$ sufficiently small. Then*

$$\frac{\partial}{\partial \varepsilon} \left((f^{\varepsilon\nu})^{-1} \overline{f^{\varepsilon\nu}} \right) \bigg|_{\varepsilon=0} = \frac{\partial}{\partial \bar{\varepsilon}} \left((f^{\varepsilon\nu})^{-1} \overline{f^{\varepsilon\nu}} \right) \bigg|_{\varepsilon=0} = 0 \tag{4.2}$$

PROOF. Since $f^{\epsilon\nu}$ is defined as a product of a holomorphic and a antiholomorphic function, it readily follows that the functions

$$\frac{\partial}{\partial \varepsilon} \left((f^{\varepsilon\nu})^{-1} \overline{f^{\varepsilon\nu}} \right) \bigg|_{\varepsilon=0}, \qquad \frac{\partial}{\partial \bar{\varepsilon}} \left((f^{\varepsilon\nu})^{-1} \overline{f^{\varepsilon\nu}} \right) \bigg|_{\varepsilon=0},$$

are harmonic, and correspond to harmonic sections of the sheaf $\mathbf{E}_{\operatorname{Ad}\rho}$, thus equal to an element in the center of $\mathfrak{g}$, which is necessarily zero since G is simple. □

Our previous construction determines a tautological real-analytic isomorphism $\mathscr{N}^s \cong \mathscr{K}^s$, and consequently, a real-analytic inclusion $\mathscr{N}^s \hookrightarrow \mathscr{K}_{\mathbb{C}}$. As we will see, the complex-analytic mechanism that we have implemented to deform a given (Γ, G)-principal bundle P_ρ can be used to extend the previous isomorphism to a real-analytic monomorphism $\mathscr{F} : T^*\mathscr{N}^s \hookrightarrow \mathscr{K}_{\mathbb{C}}$. However, the map $\mathscr{F}$ that we will construct does not correspond, in general, to the monodromy of (restricted) stable pairs [**Hit87a, Hit87b**]. The simplicity of the construction of the extension of $\mathscr{F}$ that we will provide is an indicator of its naturality: the fibers over a given $\{P_\rho\} \in \mathscr{N}^s$ are put in 1-to-1 correspondence with the classes of representations $[\chi] \in \mathscr{K}_{\mathbb{C}}$ inducing an equivalent (Γ, G)-bundle over $\mathbb{H}$.

Let us consider the holomorphic cotangent bundle $T^*\mathscr{N}^s$, whose points would be denoted as $\{P_\rho, \Phi\}$. It follows that the tangent space at a point $\{P_\rho, 0\} \in T^*\mathscr{N}$ in the zero section is isomorphic to the direct sum

$$H^1\left(\mathbb{H}, \mathbf{E}_{\operatorname{Ad}\rho}\right)^\vee \oplus H^1\left(\mathbb{H}, \mathbf{E}_{\operatorname{Ad}\rho}\right) \cong \mathfrak{S}_2(\Gamma, \operatorname{Ad}\rho) \oplus \overline{\mathfrak{S}_2(\Gamma, \operatorname{Ad}\rho)}.$$

LEMMA 3. *Let $\{\rho\} \in \mathscr{K}^s$. For every $\Phi \in \mathfrak{S}_2(\Gamma, \operatorname{Ad}\rho)$, there is a function $f_\Phi : \mathbb{H} \to G$ solving the differential equation $f_\tau f^{-1} = \Phi$, satisfying*

$$f_\Phi(\gamma\tau) = \rho(\gamma) f_\Phi(\tau) \chi(\gamma)^{-1} \qquad \forall\, \gamma \in \Gamma, \quad \tau \in \mathbb{H}, \tag{4.3}$$

for an irreducible representation $\chi : \Gamma \to G$, holomorphic and regular at each cusp. The set $\{f_\Phi\}$ of such solutions is a torsor for the right action of G.

PROOF. As in lemma 1, the existence of holomorphic solutions $f_\Phi : \mathbb{H} \to G$ of the differential equation $f_\tau f^{-1} = \Phi$, regular at the cusps, is standard [**MN84**], and a similar argument as in its proof gives, thanks to the holomorphicity of f_Φ, that the functions $\chi_\gamma(\tau) = f_\Phi(\gamma\tau)^{-1}\rho(\gamma) f_\Phi(\tau)$ are constant, and satisfy $\chi(\gamma_1\gamma_2) = \chi(\gamma_1)\chi(\gamma_2)$. Hence

$$f_\Phi(\gamma\tau) = \rho(\gamma) f_\Phi(\tau) \chi(\gamma)^{-1}, \qquad \forall\, \gamma \in \Gamma, \quad \tau \in \mathbb{H},$$

for some representation $\chi : \Gamma \to G$. Now, since f_Φ is holomorphic and regular at the cusps, it determines an isomorphism of sheaves $\mathbf{P}_\chi \cong \mathbf{P}_\rho$, implying the irreducibility of χ. Uniqueness of solutions is granted provided that a given value $f(\tau_0)$ is fixed at some $\tau_0 \in \mathbb{H}$, and hence f_Φ is defined up to right multiplication by an element in G. □

REMARK 3. The right G-action on f_Φ induces an effective $PG = G/Z(G)$-action in χ by conjugation, determining a full class $\{\chi\}$ of equivalent representations. Each function f_Φ induces a holomorphic equivalence of (Γ, G)-bundles which is invariant if ρ (and consequently Φ) is replaced by an equivalent representation, under conjugation in U. The residual right multiplication of f_Φ by elements in $Z(G)$ corresponds to the holomorphic automorphisms of P_ρ. Thus, lemma 3 determines a correspondence between the elements in $\mathfrak{S}_2(\Gamma, \operatorname{Ad}\rho)$ and the equivalence classes of irreducible representations $\{\chi\}$ whose induced (Γ, G)-bundles satisfy $P_\chi \cong P_\rho$.[4]

[4]**Note added in proof:** A result analogous to lemma 3 first appeared in the work of Florentino [**Flo01**] (lemma 2), in connection to the study of the existence problem for the so-called Schottky uniformizations of vector bundles on Riemann surfaces. Although his result is described in the case of $G = \mathrm{GL}(r, \mathbb{C})$, and purely hyperbolic Fuchsian groups, the principle and ideas that make it work are essentially the same than the ones presented here. In particular, such result is used in [**Flo01**] to characterize the set of all irreducible representations $\chi : \Gamma \to \mathrm{GL}(r, \mathbb{C})$ that induce a given equivalence class $\{E \to \Sigma\}$ of stable vector bundles on a compact Riemann surface Σ uniformized by Γ (lemma 3).

COROLLARY 2. *For every pair* $\{P_\rho, \Phi\}$, $\Phi \in \mathfrak{S}_2(\Gamma, \operatorname{Ad}\rho)$ *(up to conjugation in* U*), there is a unique induced irreducible representation* $\chi : \Gamma \to G$*, up to conjugation in* G*. The map* $\mathscr{F} : T^*\mathscr{N}^s \to \mathscr{K}_{\mathbb{C}}$, $\{P_\rho, \Phi\} \mapsto \{\chi\}$ *is 1-to-1.*

PROOF. If for $\{P_\rho, \Phi\}$, $\{P_{\rho'}, \Phi'\}$, we have that $\{\chi\} = \{\chi'\}$, then after conjugation in G, $f_{\Phi'}(f_\Phi)^{-1}$ is a regular G-valued $\rho' \otimes \rho^\vee$-automorphic function, hence a constant lying in U. Hence $\rho \cong \rho'$, or $\{\rho\} = \{\rho'\}$, which implies that $\Phi = \Phi'$ up to conjugation in U, or $\{\Phi\} = \{\Phi'\}$. □

In the next section, we will show that the differential of the map $\mathscr{F}$ is indeed a bijection at any point.

5. Shimura's isomorphism

DEFINITION 1. An *Eichler integral of weight 0 with the representation* $\operatorname{Ad}\rho$, for $\rho : \Gamma \to U$, is a holomorphic function $\mathscr{E} : \mathbb{H} \to \mathfrak{g}$ satisfying

$$\mathscr{E}(\gamma\tau) = \operatorname{Ad}\rho(\gamma) \cdot \mathscr{E}(\tau) + z(\gamma), \quad \gamma \in \Gamma, \tau \in \mathbb{H},$$

where $z : \Gamma \to \mathfrak{g}$ is an element in $Z^1_P(\Gamma, \mathfrak{g}_{\operatorname{Ad}\rho})$, and moreover, such that $\mathscr{E}'$ is a $\mathfrak{g}$-valued automorphic form of weight 2 for $\operatorname{Ad}\rho$, which is meromorphic at each parabolic fixed point of Γ.

Parabolic cohomology is related to cusp forms of weight 2 by means of *Shimura's isomorphism* [**Shi71**]

$$\mathfrak{S}_2(\Gamma, \operatorname{Ad}\rho) \cong H^1_P(\Gamma, \mathfrak{u}_{\operatorname{Ad}\rho}),$$

for $\mathfrak{u}$-valued cocycles with the representation $\operatorname{Ad}\rho$, given explicitly by

$$\mathscr{L}_S(\Phi) = \text{the cohomology class of } \operatorname{Re}(z_\Phi)$$

where $z_\Phi : \Gamma \to \mathfrak{g}$ is any 1-cocycle associated to an antiderivative of Φ (an Eichler integral of weight 0), so that $\operatorname{Re}(z_\Phi) \in Z^1_P(\Gamma, \mathfrak{u}_{\operatorname{Ad}\rho})$.

Shimura's isomorphism is an isomorphism of real vector spaces. Since $\mathfrak{g} \cong \mathfrak{u}_{\mathbb{C}}$ implies the canonical identification $H^1_P(\Gamma, \mathfrak{g}_{\operatorname{Ad}\rho}) \cong H^1_P(\Gamma, \mathfrak{u}_{\operatorname{Ad}\rho})_{\mathbb{C}}$, and recalling that for a complex vector space V, $(V_{\mathbb{R}})_{\mathbb{C}} \cong V \oplus \overline{V}$, it is also possible to formulate Shimura's isomorphism as an isomorphism of complex vector spaces, in the form

$$\mathfrak{S}_2(\Gamma, \operatorname{Ad}\rho) \oplus \overline{\mathfrak{S}_2(\Gamma, \operatorname{Ad}\rho)} \cong H^1_P(\Gamma, \mathfrak{g}_{\operatorname{Ad}\rho}).$$

For fixed $\Phi \in \mathfrak{S}_2(\Gamma, \operatorname{Ad}\rho)$ and $\nu \in \overline{\mathfrak{S}_2(\Gamma, \operatorname{Ad}\rho)}$, let us denote

$$\dot{f}_+ = \left.\frac{\partial f_{\epsilon\Phi}}{\partial \epsilon}\right|_{\epsilon=0}, \qquad \dot{f}^- = \left.\frac{\partial f^{\epsilon\nu}}{\partial \bar{\epsilon}}\right|_{\epsilon=0}.$$

LEMMA 4. *For any* $\Phi \in \mathfrak{S}_2(\Gamma, \operatorname{Ad}\rho)$, $\nu \in \overline{\mathfrak{S}_2(\Gamma, \operatorname{Ad}\rho)}$ *and* $\epsilon \in \mathbb{C}$ *sufficiently small, the functions* $\dot{f}_+$ *and* $\dot{f}^-$ *are Eichler integrals of weight 0, satisfying*

$$(\dot{f}_+)' = \Phi, \qquad (\dot{f}^-)' = \overline{\nu}. \tag{5.1}$$

PROOF. That $\dot{f}_+$ is an antiderivative of Φ is automatic from (4.3) and the holomorphicity of $f_{\epsilon\Phi}$. The argument for $\dot{f}^-$ is slightly more elaborate: first of all, it follows from (4.1) that $(\dot{f}^-)_{\bar{\tau}} = 0$, so $\dot{f}^-$ is holomorphic. Now, it follows from lemma 2 that

$$\overline{\left.\frac{\partial f^{\epsilon\nu}}{\partial \epsilon}\right|_{\epsilon=0}} = \left.\frac{\partial f^{\epsilon\nu}}{\partial \bar{\epsilon}}\right|_{\epsilon=0}$$

but the left-hand side is an antiderivative for $\overline{\nu}$ as a consequence of (4.1), and the claim follows. $\square$

Proof of theorem 1. We have seen in section 4 that the tangent space at any point $\{P_\rho, 0\} \in T^*\mathscr{N}^s$ is modeled by the space $\mathfrak{S}_2(\Gamma, \operatorname{Ad}\rho) \oplus \overline{\mathfrak{S}_2(\Gamma, \operatorname{Ad}\rho)}$. In turn, the tangent space at $\{\rho\} = \mathscr{F}(\{P_\rho, 0\}) \in \mathscr{K}_{\mathbb{C}}$ is $H^1_P(\Gamma, \mathfrak{g}_{\operatorname{Ad}\rho})$.

The isomorphism $(V_{\mathbb{R}})_{\mathbb{C}} \cong V \oplus \overline{V}$ is determined by the $\sqrt{-1}$ and $-\sqrt{-1}$-eigenspaces of the map $(v, w) \mapsto (-w, v)$, $v, w \in V$, which are complex subspaces of $(V_{\mathbb{R}})_{\mathbb{C}}$ spanned, respectively, by elements of the form $(v, -\sqrt{-1}v)$ and $(v, \sqrt{-1}v)$. Under this identification, the isomorphism

$$\mathscr{L}_S : \mathfrak{S}_2(\Gamma, \operatorname{Ad}\rho) \oplus \overline{\mathfrak{S}_2(\Gamma, \operatorname{Ad}\rho)} \to H^1_P(\Gamma, \mathfrak{g}_{\operatorname{Ad}\rho})$$

is given explicitly as

$$\mathscr{L}_S(\Psi, \nu) = [z_\Psi + \overline{z_{\overline{\nu}}}], \tag{5.2}$$

where $[\cdot]$ denotes the cohomology class of an element in $Z^1(\Gamma, \mathfrak{g}_{\operatorname{Ad}\rho})$.

Choose an arbitrary representative ρ in its equivalence class, and let ϵ be sufficiently small, and $\chi_{\epsilon\Psi}$ (resp. $\{\rho^{\epsilon\nu}\}$) the associated irreducible representation in $\mathscr{K}_{\mathbb{C}}$ (resp. $\mathscr{K}^s$). Then, the maps $\{\epsilon\Psi\} \mapsto \{\chi_{\epsilon\Psi}\}$, $\{\epsilon\nu\} \mapsto \{\rho^{\epsilon\nu}\}$ defined in corollaries 1 and 2 yield

$$d\mathscr{F}_{\{P_\rho,0\}}(\Psi) = -\frac{\partial}{\partial\epsilon}(\chi_{\epsilon\Psi})\bigg|_{\epsilon=0} \rho^{-1}, \qquad d\mathscr{F}_{\{P_\rho,0\}}(\nu) = \frac{\partial}{\partial\epsilon}(\rho^{\epsilon\nu})\bigg|_{\epsilon=0} \rho^{-1},$$

as $\mathfrak{g}$-valued parabolic 1-cocycles. On the other hand, lemma 4 states that $\dot{f}_+$ is an Eichler integral of weight 0 with the representation $\operatorname{Ad}\rho$, satisfying $(\dot{f}_+)' = \Psi$. Its associated parabolic 1-cocycle is seen to be

$$z_+(\gamma) = -\frac{\partial}{\partial\epsilon}(\chi_{\epsilon\Psi}(\gamma))\bigg|_{\epsilon=0} \rho(\gamma)^{-1}.$$

In turn, $\dot{f}^-$ is also an Eichler integral of weight 0, satisfying $(\dot{f}^-)' = \overline{\nu}$. From (4.1) and (4.2), its associated parabolic 1-cocycle is

$$z^-(\gamma) = \frac{\partial}{\partial\overline{\epsilon}}(\rho^{\epsilon\nu}(\gamma))\bigg|_{\epsilon=0} \rho(\gamma)^{-1} = \overline{\frac{\partial}{\partial\epsilon}(\rho^{\epsilon\nu}(\gamma))\bigg|_{\epsilon=0} \rho(\gamma)^{-1}},$$

where the last equality follows from the algebraic identity $(\rho^{\epsilon\nu})^{-1}\overline{\rho^{\epsilon\nu}} = I$, since $\mathfrak{u}$ is the subspace of $\mathfrak{g}$ of elements X satisfying $\overline{X} = X$.

Finally, since the conjugation of ρ induces a trivial change in the parabolic 1-cocycles, leaving their cohomology class fixed, we conclude

$$d\mathscr{F}_{\{P_\rho,0\}}(\Psi, 0) = \mathscr{L}_S(\Psi, 0), \qquad d\mathscr{F}_{\{P_\rho,0\}}(0, \nu) = \mathscr{L}_S(0, \nu),$$

which completes the proof. $\square$

In particular, theorem 1 ensures that the restriction of $\mathscr{F}$ to the image of s_0 is a real-analytic diffeomorphism onto $\mathscr{K}^s$. Let $\pi : T^*\mathscr{N}^s \to \mathscr{N}^s$ be the cotangent bundle projection. In general, the vertical subbundle $V \to T^*\mathscr{N}^s$ of $T(T^*\mathscr{N}^s)$, defined as $\ker(d\pi)$, is canonically isomorphic to $\pi^*(T^*\mathscr{N}^s)$, and there is an induced exact sequence of vector bundles over $T^*\mathscr{N}^s$,

$$0 \to \pi^*(T^*\mathscr{N}^s) \to T(T^*\mathscr{N}^s) \to \pi^*(T\mathscr{N}^s) \to 0,$$

although, for an arbitrary $\Phi \in \mathfrak{S}_2(\Gamma, \operatorname{Ad}\rho)$, there is no canonical splitting of $T_{\{P_\rho,\Phi\}}(T^*\mathscr{N}^s)$, and consequently no natural generalization of the result in the complement of s_0 on $T^*\mathscr{N}^s$. Nevertheless, it readily follows that, indeed, the differential of $\mathscr{F}$ is a bijection over any point $\{P_\rho, \Phi\} \in T^*\mathscr{N}^s$. To see this, consider the map $\mathrm{pr} : \mathscr{F}(T^*\mathscr{N}^s) \subset \mathscr{K}_\mathbb{C} \to \mathscr{K}^s$, defined by requiring

$$\mathscr{F}_U \circ \pi = \mathrm{pr} \circ \mathscr{F},$$

where $\mathscr{F}_U = \mathscr{F} \circ s_0 : \mathscr{N}^s \to \mathscr{K}^s$. We know that $d\mathscr{F}_U$ is an isomorphism at any point in $\mathscr{N}^s$, and consequently, $d\left(\mathscr{F}_U \circ \pi\right)$ is surjective at any $\{P_\rho, \Phi\} \in T^*\mathscr{N}^s$, and therefore, the same is true for $d\left(\mathrm{pr} \circ \mathscr{F}\right)$. From this, it follows that the kernel of $d\mathscr{F}$ at a point $\{P_\rho, \Phi\} \in T^*\mathscr{N}^s$ must be a subspace of $V_{\{P_\rho,\Phi\}} = \ker\left(d\pi_{\{P_\rho,\Phi\}}\right)$. On the other hand, the isomorphism $V \cong \pi^*\left(T^*\mathscr{N}^s\right)$ implies that $d\mathscr{F}|_V$ is given at any point $\{P_\rho, \Phi\}$ simply as

$$\Psi \mapsto \mathscr{L}_S(\Psi, 0).$$

Hence $d\mathscr{F}\left(V_{\{P_\rho,\Phi\}}\right)$ has maximal rank, and we have concluded the following corollary.

COROLLARY 3. *The map*

$$\mathscr{F} : T^*\mathscr{N}^s \to \mathscr{K}_\mathbb{C},$$

is a real-analytic monomorphism. Its restriction to the zero section is in correspondence with the inclusion $\iota : \mathscr{K}^s \hookrightarrow \mathscr{K}_\mathbb{C}$. *The image* $\mathscr{F}\left(T^*\mathscr{N}^s\right) = \mathscr{K}_\mathbb{C}^s$ *is the Zariski open set in* $\mathscr{K}_\mathbb{C}$ *of equivalence classes of* χ *whose associated* (Γ, G)*-bundle is isomorphic to an irreducible unitary one, without nontrivial automorphisms. There is a foliation of* $\mathscr{K}_\mathbb{C}^s$, *parametrized by* $\mathscr{K}^s$, *whose leaves are given by the images of each* $T^*_{\{P_\rho\}}\mathscr{N}^s \cong \mathfrak{S}_2(\Gamma, \mathrm{Ad}\,\rho)$ *under* $\mathscr{F}$.

REMARK 4. Altogether, the real isomorphisms $\mathfrak{S}_2(\Gamma, \mathrm{Ad}\,\rho) \cong H^1_P(\Gamma, \mathfrak{u}_{\mathrm{Ad}\,\rho})$ turn the smooth locus of the real character variety $\mathscr{K}$ into an almost complex manifold. Theorem 1, restricted to $\mathscr{N}^s$, shows indirectly that this almost complex structure on $\mathscr{K}^s$ would be integrable (in the same spirit of [**NS64**]), thus determining a complex structure over $\mathscr{K}^s$.

REMARK 5. Using holomorphic coordinates, the Petersson inner product in $\overline{\mathfrak{S}_2(\Gamma, \mathrm{Ad}\,\rho)}$ induces a Kähler stucture in $\mathscr{N}^s$, given as

$$\Omega\left(\frac{\partial}{\partial\epsilon(\nu_1)}, \overline{\frac{\partial}{\partial\epsilon(\nu_2)}}\right) = \frac{\sqrt{-1}}{2}\langle\nu_1, \nu_2\rangle_P.$$

Its imaginary part is a closed, real 2-form defining a symplectic structure in $\mathscr{N}^s$.

Since cup product in parabolic cohomology corresponds to wedge product (defined in terms of the Killing form) in the space of cohomology classes of $\mathrm{Ad}\,\rho$-automorphic 1-forms on $\mathbb{H}$ with values in $\mathfrak{u}$ and compactly supported within a fundamental region of Γ, a space in turn isomorphic to $H^1(\mathbb{H}, \mathbf{E}_{\mathrm{Ad}\,\rho})$ (cf. [**NS64**], Proposition 4.4), Shimura's isomorphism is an isomorphism of symplectic vector spaces, for the symplectic form on $H^1_P(\Gamma, \mathfrak{u}_{\mathrm{Ad}\,\rho})$ given as

$$H^1_P(\Gamma, \mathfrak{u}_{\mathrm{Ad}\,\rho}) \times H^1_P(\Gamma, \mathfrak{u}_{\mathrm{Ad}\,\rho}) \to H^2(\Gamma, \mathbb{R}) \cong \mathbb{R},$$

$$(u_1, u_2) \mapsto (u_1 \smile u_2)(\gamma_1, \gamma_2) = -K\left(u_1(\gamma_1), \mathrm{Ad}\,\rho(\gamma_1)u_2(\gamma_2)\right).$$

The latter defines Goldman's symplectic structure in $\mathscr{K}^s$ [**Gol84**]. Thus, the map $\mathscr{F}_U$ is also a symplectomorphism between $\mathscr{N}^s$ and $\mathscr{K}^s$.

References

[BS15] V. Balaji and C. S. Seshadri, *Moduli of parahoric $\mathcal{G}$-torsors on a compact Riemann surface*, J. Algebraic Geom. **24** (2015), no. 1, 1–49. MR3275653

[Flo01] C. Florentino, *Schottky uniformization and vector bundles over Riemann surfaces*, Manuscripta Math. **105** (2001), no. 1, 69–83. MR1885814

[Gol84] W. M. Goldman, *The symplectic nature of fundamental groups of surfaces*, Adv. in Math. **54** (1984), no. 2, 200–225. MR762512

[Hej83] D. A. Hejhal, *The Selberg trace formula for* PSL(2, **R**). *Vol. 2*, Lecture Notes in Mathematics, vol. 1001, Springer-Verlag, Berlin, 1983. MR711197

[Hel79] S. Helgason, *Differential geometry, Lie groups, and symmetric spaces*, Pure and Applied Mathematics, vol. 80, Academic Press, Inc. [Harcourt Brace Jovanovich, Publishers], New York-London, 1978. MR514561

[Hit87a] N. J. Hitchin, *The self-duality equations on a Riemann surface*, Proc. London Math. Soc. (3) **55** (1987), no. 1, 59–126. MR887284

[Hit87b] N. Hitchin, *Stable bundles and integrable systems*, Duke Math. J. **54** (1987), no. 1, 91–114. MR885778

[Iva88] N. V. Ivanov, *Projective structures, flat bundles and Kähler metrics on moduli spaces* (Russian), Mat. Sb. (N.S.) **133(175)** (1987), no. 2, 208–222, 272; English transl., Math. USSR-Sb. **61** (1988), no. 1, 211–224. MR905006

[Leh64] J. Lehner, *Discontinuous groups and automorphic functions*, Mathematical Surveys, No. VIII, American Mathematical Society, Providence, R.I., 1964. MR0164033

[Men17] C. Meneses, *On vector-valued Poincaré series of weight 2*, J. Geom. Phys. **120** (2017), 317–329. MR3712164

[MN84] B. Malgrange, *Lectures on the theory of functions of several complex variables*, Tata Institute of Fundamental Research Lectures on Mathematics and Physics, vol. 13, Distributed for the Tata Institute of Fundamental Research, Bombay; by Springer-Verlag, Berlin, 1984. Reprint of the 1958 edition; Notes by Raghavan Narasimhan. MR742775

[MS80] V. B. Mehta and C. S. Seshadri, *Moduli of vector bundles on curves with parabolic structures*, Math. Ann. **248** (1980), no. 3, 205–239. MR575939

[NS64] M. S. Narasimhan and C. S. Seshadri, *Holomorphic vector bundles on a compact Riemann surface*, Math. Ann. **155** (1964), 69–80. MR0166799

[NS65] M. S. Narasimhan and C. S. Seshadri, *Stable and unitary vector bundles on a compact Riemann surface*, Ann. of Math. (2) **82** (1965), 540–567. MR0184252

[Pet40] H. Petersson, *Über eine Metrisierung der automorphen Formen und die Theorie der Poincaréschen Reihen* (German), Math. Ann. **117** (1940), 453–537. MR0002624

[Pet48] H. Petersson, *Automorphe Formen als metrische Invarianten. I. Automorphe Formen, metrische Verknüpfung, Eigenfunktionen linearer Funktionale* (German), Math. Nachr. **1** (1948), 158–212. MR0027860

[Ram75] A. Ramanathan, *Stable principal bundles on a compact Riemann surface*, Math. Ann. **213** (1975), 129–152. MR0369747

[Ses10] C. S. Seshadri, *Moduli of π-vector bundles over an algebraic curve*, Questions on Algebraic Varieties (C.I.M.E., III Ciclo, Varenna, 1969), Edizioni Cremonese, Rome, 1970, pp. 139–260. MR0280496

[Shi71] G. Shimura, *Introduction to the arithmetic theory of automorphic forms*, Princeton University Press, 1971.

[TZ08] L. A. Takhtajan and P. Zograf, *The first Chern form on moduli of parabolic bundles*, Math. Ann. **341** (2008), no. 1, 113–135. MR2377472

[Wei64] A. Weil, *Remarks on the cohomology of groups*, Ann. of Math. (2) **80** (1964), 149–157. MR0169956

Max Planck Institut für Mathematik, Vivatsgasse 7, 53111 Bonn, Germany

Current address: CIMAT A.C., Jalisco S/N, Col. Valenciana CP: 36240, Gto., Mexico

Email address: `claudio@mpim-bonn.mpg.de, claudio.meneses@cimat.mx`

Contemporary Mathematics
Volume **709**, 2018
http://dx.doi.org/10.1090/conm/709/14296

On the dual of BV

Monica Torres

ABSTRACT. An important problem in geometric measure theory is the characterization of BV^*, the dual of the space of functions of bounded variation. In this paper we survey recent results that show that the solvability of the equation $div\,\boldsymbol{F} = T$ is closely connected to the problem of characterizing BV^*. In particular, the (signed) measures in BV^* can be characterized in terms of the solvability of the equation $div\,\boldsymbol{F} = T$.

1. Introduction

It is an open problem in geometric measure theory to give a full characterization of BV^*, the dual of the space of functions of bounded variation (see Ambrosio-Fusco-Pallara [**AFP**, Remark 3.12]). Meyers and Ziemer characterized in [**MZ**] the positive measures in $\mathbb{R}^n$ that belong to the dual of $BV(\mathbb{R}^n)$, where $BV(\mathbb{R}^n)$ is the space of all functions in $L^1(\mathbb{R}^n)$ whose distributional gradient is a vector-measure in $\mathbb{R}^n$ with finite total variation. They showed that the positive measure μ belongs to $BV(\mathbb{R}^n)^*$ if and only if μ satisfies the condition

$$\mu(B(x,r)) \leq Cr^{n-1}$$

for every open ball $B(x,r) \subset \mathbb{R}^n$ and $C = C(n)$. Besides the classical paper by Meyers and Ziemer, we refer the interested reader to the paper by De Pauw [**DP**], where the author analyzes SBV^*, the dual of the space of special functions of bounded variation (see also Fusco-Spector [**FS**] for an integral representation of elements in the dual of BV, under the assumption of the continuum hypothesis).

In Phuc-Torres [**PT**] we showed that there is a connection between the problem of characterizing BV^* and the solvability of the equation $div\,\boldsymbol{F} = T$. Indeed, we showed that the (signed) measure μ belongs to $BV(\mathbb{R}^n)^*$ if and only if there exists a bounded vector field $\boldsymbol{F} \in L^\infty(\mathbb{R}^n, \mathbb{R}^n)$ such that $div\,\boldsymbol{F} = \mu$. Also, we showed that μ belongs to $BV(\mathbb{R}^n)^*$ if and only if

$$|\mu(U)| \leq C\,\mathcal{H}^{n-1}(\partial U) \tag{1.1}$$

for any open (or closed) set $U \subset \mathbb{R}^n$ with smooth boundary.

In De Pauw-Torres [**DPT**], another BV-type space was considered, the space $BV_{\frac{n}{n-1}}(\mathbb{R}^n)$, defined as the space of all functions $u \in L^{\frac{n}{n-1}}(\mathbb{R}^n)$ such that Du, the distributional gradient of u, is a vector-measure in $\mathbb{R}^n$ with finite total variation.

2010 *Mathematics Subject Classification.* Primary 46E27, 26A45, 49Q15.
Key words and phrases. Dual of BV, image processing, measures.
The author was supported in part by the Simons Foundation Grant #524190.

A closed subspace of $BV_{\frac{n}{n-1}}(\mathbb{R}^n)^*$, which is a Banach space denoted as CH_0, was characterized in [**DPT**] and it was proven that $T \in CH_0$ if and only if $T = div\,\boldsymbol{F}$, for a continuous vector field $\boldsymbol{F}$ vanishing at infinity.

In Phuc-Torres [**PT17**], the analysis of $BV(\mathbb{R}^n)^*$ and $BV_{\frac{n}{n-1}}(\mathbb{R}^n)^*$ was continued. It was shown that $BV(\mathbb{R}^n)^*$ and $BV_{\frac{n}{n-1}}(\mathbb{R}^n)^*$ are isometrically isomorphic. It was also shown that the measures in $BV_{\frac{n}{n-1}}(\mathbb{R}^n)^*$ coincide with the measures in $\dot{W}^{1,1}(\mathbb{R}^n)^*$, the dual of the homogeneous Sobolev space $\dot{W}^{1,1}(\mathbb{R}^n)$, in the sense of isometric isomorphism. We remark that the space $\dot{W}^{1,1}(\mathbb{R}^n)^*$ is denoted as the G space in image processing (see Meyer [**ME**]), and that it plays a key role in modeling the noise of an image. Thus, the results in [**PT17**] provide a formula to compute the G-norm of a measure (see Remark 3.7) that is more suitable for computations.

It is obvious that if μ is a locally finite signed Radon measure then $\|\mu\| \in BV(\mathbb{R}^n)^*$ implies that $\mu \in BV(\mathbb{R}^n)^*$. The converse was unknown to Meyers and Ziemer as they raised this issue in their classical paper [**MZ**, page 1356]. In [**PT17**], we showed that the converse does not hold true in general by constructing a locally integrable function f such that $f \in BV(\mathbb{R}^n)^*$ but $|f| \notin BV(\mathbb{R}^n)^*$.

Given a bounded open set Ω with Lipschitz boundary, $BV_0(\Omega)$ is the space of functions of bounded variation with zero trace on $\partial\Omega$. All (signed) measures in Ω that belong to $BV_0(\Omega)^*$ were characterized in [**PT17**]. It was shown that a locally finite signed measure μ belongs to $BV_0(\Omega)^*$ if and only if (1.1) holds for any smooth open (or closed) set $U \subset\subset \Omega$, and if and only if $\mu = div\,\boldsymbol{F}$ for a vector field $\boldsymbol{F} \in L^\infty(\Omega, \mathbb{R}^n)$. Moreover, the measures in $BV_0(\Omega)^*$ coincide with the measures in $W_0^{1,1}(\Omega)^*$, in the sense of isometric isomorphism.

In the case of $BV(\Omega)$, the space of functions of bounded variation in a bounded open set Ω with Lipschitz boundary (but without the condition of having zero trace on $\partial\Omega$), we shall restrict our attention only to measures in $BV(\Omega)^*$ with bounded total variation in Ω, i.e., finite measures. This is in a sense natural since any *positive* measure that belongs to $BV(\Omega)^*$ must be finite due to the fact that the function $1 \in BV(\Omega)$. We showed in [**PT17**] that a finite measure μ belongs to $BV(\Omega)^*$ if and only if (1.1) holds for every smooth open set $U \subset\subset \mathbb{R}^n$, where μ is extended by zero to $\mathbb{R}^n \setminus \Omega$.

The solvability of the equation $div\,\boldsymbol{F} = T$, in various spaces of functions, has been studied in Bourgain-Brezis [**BB**], De Pauw-Pfeffer [**DPP**], De Pauw-Torres [**DPT**], Phuc-Torres [**PT**] (see also Tadmor [**T**], Bousquet-Mironescu-Russ [**BMR**], Russ [**R**] and the references therein). A first attempt to solve the equation $div\,\boldsymbol{F} = T$ can be made by solving the Laplace's equation. Thus, we can start by setting $\boldsymbol{F} = \nabla u$ and solving $\Delta u = f \in L^p(U)$, $1 < p < \infty$. In this case, there exists a solution $u \in W^{2,p}(U)$ and hence $\boldsymbol{F} \in W^{1,p}(U)$. However, the limiting cases $p = 1$ and $p = \infty$ can not be solved in general. Indeed, McMullen [**M**] and Preiss [**P**] have shown that there exist functions $f \in L^\infty$ (even continuous f) such that the equation $div\,\boldsymbol{F} = f$ has no Lipschitz solution (see also Dacorogna-Fusco-Tartar [**DFT**] and Bourgain-Brezis [**BB**, Section 2.2]). Analogously, there exist functions $f \in L^1$ such that the equation $div\,\boldsymbol{F} = f$ has no solution in BV and not even in $L^{n/n-1}$ (see Bourgain-Brezis [**BB**, Section 2.1]).

We recall that if u solves $\Delta u = f \in L^p(U)$ and $n<p<\infty$ then $\nabla u \in C^{0,1-\frac{n}{p}}(U)$. From here we see that this regularity fails for the limiting case $p=\infty$, that would correspond to having a Lipschitz solution, and for the case $p = n$, that would correspond to having a continuous solution. For the case $p = n$, since the equation $\Delta u = f \in L^n$ admits a solution $u \in W^{2,n}$, it follows that $div\boldsymbol{F} = f \in L^n$ has a solution $\boldsymbol{F} \in W^{1,n}$. However, $W^{1,n}$ is not contained in L^∞ (since this is a limiting case in the Sobolev embedding theorem) and hence one can not conclude that $\boldsymbol{F} \in L^\infty$. The previous discussion suggests to study the limiting case $f \in L^n$ and to research the existence of vector fields $\boldsymbol{F} \in L^\infty$ or $\boldsymbol{F}$ continuous that solve equation $div\boldsymbol{F} = f \in L^n$. Nirenberg provided the example $u(x) = \varphi(x)x_1 \left|\log \|x\|\right|^\alpha$, $0 < \alpha < \frac{n-1}{n}$, where φ is smooth supported near 0, which shows that solving the equation for the case $p = n$ requires methods other than the Laplace's equation. Indeed, one can check for this particular u that $\Delta u \in L^n(\mathbb{R}^n)$ but $\nabla u \notin L^\infty_{loc}(\mathbb{R}^n)$.

The work by Bourgain-Brezis [**BB**] initiated the analysis of the limiting case $p = n$. The equation $div\boldsymbol{F} = f \in L^n_\#([0,1]^n)$ was solved in [**BB**] in the class $L^\infty([0,1]^n, \mathbb{R}^n) \cap W^{1,n}([0,1]^n; \mathbb{R}^n)$. The solvability of $div\boldsymbol{F} = T$ in the space of continuous vector fields $C(U, \mathbb{R}^n)$ was studied by De Pauw-Pfeffer [**DPP**]. They showed that the equation has a continuous solution if and only if T belongs to a class of distributions, the space of strong charges, denoted as $CH_n(U)$. Moreover, $L^n(U) \subset CH_n(U)$, solving in particular the critical case $div\boldsymbol{F} = f \in L^n$. The case $p = n$ was also addressed in De Pauw-Torres [**DPT**] and it is discussed in Section 5 of this survey paper.

The organization of this paper is as follows. In Section 2 we introduce the space of functions of bounded variation BV. In Section 3 we present the characterization of all (signed) measures in $BV_{\frac{n}{n-1}}(\mathbb{R}^n)^*$. In Section 4 we discuss a counterexample that resolves a question raised by Meyers and Ziemer in [**MZ**, page 1356]. In Section 5 we characterize a subspace of $BV_{\frac{n}{n-1}}(\mathbb{R}^n)^*$ in terms of the solvability of $div\boldsymbol{F} = T$, in the class of continuous vector fields vanishing at infinity. Finally, in Section 6 we discuss the characterization of all (signed) measures in $BV_0(\Omega)^*$, and all finite measures in the space $BV(\Omega)^*$.

2. Functions of bounded variation

In this section we define all the spaces that will be relevant in this paper.

DEFINITION 2.1. Let Ω be any open set. The space $\mathcal{M}(\Omega)$ consists of all finite (signed) Radon measures μ in Ω; that is, the total variation of μ, denoted as $\|\mu\|$, satisfies $\|\mu\|(\Omega) < \infty$. The space $\mathcal{M}_{loc}(\Omega)$ consists of all locally finite Radon measures μ in Ω; that is, $\|\mu\|(K) < \infty$ for every compact set $K \subset \Omega$.

Note here that $\mathcal{M}_{loc}(\Omega)$ is identified with the dual of the locally convex space $C_c(\Omega)$ (the space of continuous real-valued functions with compact support in Ω) (see [**D**]), and thus it is a real vector space. For $\mu \in \mathcal{M}_{loc}(\Omega)$, it is not required that either the positive part or the negative part of μ has finite total variation linebreak in Ω.

In the next definition by a *vector-valued measure* we mean a Radon measure that takes values in $\mathbb{R}^n$.

DEFINITION 2.2. Let Ω be any open set. The space of functions of bounded variation, denoted as $BV(\Omega)$, is defined as the space of all functions $u \in L^1(\Omega)$ such that the distributional gradient Du is a finite vector-valued measure in Ω. For $\Omega \neq \mathbb{R}^n$, we equip $BV(\Omega)$ with the norm

$$\|u\|_{BV(\Omega)} = \|u\|_{L^1(\Omega)} + \|Du\|(\Omega), \tag{2.1}$$

where $\|Du\|(\Omega)$ denotes the total variation of the vector-valued measure Du over Ω. For $\Omega = \mathbb{R}^n$, following Meyers-Ziemer [**MZ**], we will instead equip $BV(\mathbb{R}^n)$ with the homogeneous norm given by

$$\|u\|_{BV(\mathbb{R}^n)} = \|Du\|(\mathbb{R}^n). \tag{2.2}$$

Another BV-like space is $BV_{\frac{n}{n-1}}(\mathbb{R}^n)$, defined as the space of all functions in $L^{\frac{n}{n-1}}(\mathbb{R}^n)$ such that Du is a finite vector-valued measure. The space $BV_{\frac{n}{n-1}}(\mathbb{R}^n)$ is a Banach space when equipped with the norm

$$\|u\|_{BV_{\frac{n}{n-1}}(\mathbb{R}^n)} = \|Du\|(\mathbb{R}^n).$$

REMARK 2.3. By definition $BV(\mathbb{R}^n) \subset L^1(\mathbb{R}^n)$ and thus it is a normed space under the norm (2.2). However, $BV(\mathbb{R}^n)$ is not complete under this norm. Also, we have

$$\|Du\|(\Omega) = \sup\left\{\int_\Omega u \operatorname{div}\varphi dx : \varphi \in C_c^1(\Omega) \text{ and } |\varphi(x)| \leq 1 \, \forall x \in \Omega\right\},$$

where $\varphi = (\varphi_1, \varphi_2, ..., \varphi_n)$ and $|\varphi(x)| = (\varphi_1(x)^2 + \varphi_2(x)^2 + \cdots + \varphi_n(x)^2)^{1/2}$. In what follows, we shall also write $\int_\Omega |Du|$ instead of $\|Du\|(\Omega)$.

We recall the following Sobolev's inequality for functions in $BV(\mathbb{R}^n)$ whose proof can be found in [**AFP**, Theorem 3.47]:

$$\|u\|_{L^{\frac{n}{n-1}}(\mathbb{R}^n)} \leq C(n) \|Du\|(\mathbb{R}^n), \quad u \in BV(\mathbb{R}^n). \tag{2.3}$$

Inequality (2.3) immediately implies the following continuous embedding

$$BV(\mathbb{R}^n) \hookrightarrow BV_{\frac{n}{n-1}}(\mathbb{R}^n). \tag{2.4}$$

We recall that the standard Sobolev space $W^{1,1}(\Omega)$ is defined as the space of all functions $u \in L^1(\Omega)$ such that $Du \in L^1(\Omega)$. The Sobolev space $W^{1,1}(\Omega)$ is a Banach space with the norm

$$\begin{aligned} \|u\|_{W^{1,1}(\Omega)} &= \|u\|_{L^1(\Omega)} + \|Du\|_{L^1(\Omega)} \\ &= \int_\Omega \left[|u| + (|D_1u|^2 + |D_2u|^2 + \cdots + |D_nu|^2)^{\frac{1}{2}}\right] dx. \end{aligned} \tag{2.5}$$

Hereafter, we let $C_c^\infty(\Omega)$ denote the space of smooth functions with compact support in a general open set Ω. We will often refer to the following homogeneous Sobolev space.

DEFINITION 2.4. Let $\dot{W}^{1,1}(\mathbb{R}^n)$ denote the space of all functions $u \in L^{\frac{n}{n-1}}(\mathbb{R}^n)$ such that $Du \in L^1(\mathbb{R}^n)$. Equivalently, the space $\dot{W}^{1,1}(\mathbb{R}^n)$ can also be defined as the closure of $C_c^\infty(\mathbb{R}^n)$ in $BV_{\frac{n}{n-1}}(\mathbb{R}^n)$ (i.e., in the norm $\|Du\|_{L^1(\mathbb{R}^n)}$). Thus, $u \in \dot{W}^{1,1}(\mathbb{R}^n)$ if and only if there exists a sequence $u_k \in C_c^\infty(\mathbb{R}^n)$ such that $\int_{\mathbb{R}^n} |D(u_k - u)| dx = 0$, and moreover,

$$\dot{W}^{1,1}(\mathbb{R}^n) \hookrightarrow BV_{\frac{n}{n-1}}(\mathbb{R}^n).$$

We recall that the open bounded set Ω has Lipschitz boundary if for each $x \in \partial\Omega$, there exist $r > 0$ and a Lipschitz mapping $h : \mathbb{R}^{n-1} \to \mathbb{R}$ such that, upon rotating and relabeling the coordinate axes if necessary, we have

$$\Omega \cap B(x,r) = \{y = (y_1, \ldots, y_{n-1}, y_n) : h(y_1, \ldots, y_{n-1}) < y_n\} \cap B(x,r).$$

REMARK 2.5. Let Ω be a bounded open set with Lipschitz boundary. We denote by $W_0^{1,1}(\Omega)$ the Sobolev space consisting of all functions in $W^{1,1}(\Omega)$ with zero trace on $\partial\Omega$. Then it is well-known that $C_c^\infty(\Omega)$ is dense in $W_0^{1,1}(\Omega)$. We will explain in Section 6 the precise definition of $BV_0(\Omega)$, the space of all functions in $BV(\Omega)$ with zero trace on $\partial\Omega$ (see (6.1)). In this paper we equip the two spaces, $BV_0(\Omega)$ and $W_0^{1,1}(\Omega)$, with the equivalent norms (see (6.2)) to (2.1) and (2.5), respectively, given by

$$\|u\|_{BV_0(\Omega)} = \|Du\|(\Omega), \quad \text{and} \quad \|u\|_{W_0^{1,1}(\Omega)} = \int_\Omega |Du| dx.$$

DEFINITION 2.6. For any open set Ω, we let $BV_c(\Omega)$ denote the space of functions in $BV(\Omega)$ with compact support in Ω. Also, $BV^\infty(\Omega)$ and $BV_0^\infty(\Omega)$ denote the space of bounded functions in $BV(\Omega)$ and $BV_0(\Omega)$, respectively. Finally, $BV_c^\infty(\Omega)$ is the space of all bounded functions in $BV(\Omega)$ with compact support in Ω.

3. Characterization of (signed) measures in $BV_{\frac{n}{n-1}}(\mathbb{R}^n)^*$

The fundamental step in the characterization of the (signed) measures in $BV_{\frac{n}{n-1}}(\mathbb{R}^n)^*$ is the fact that $BV_c^\infty(\mathbb{R}^n)$ is dense in $BV_{\frac{n}{n-1}}(\mathbb{R}^n)$. Indeed, we have the following ([**PT17**, Theorem 3.1]:

THEOREM 3.1. *Let $u \in BV_{\frac{n}{n-1}}(\mathbb{R}^n)$, $u \geq 0$, and $\phi_k \in C_c^\infty(\mathbb{R}^n)$ be a nondecreasing sequence of smooth functions satisfying:*

$$0 \leq \phi_k \leq 1,\ \phi_k \equiv 1 \text{ on } B_k(0),\ \phi_k \equiv 0 \text{ on } \mathbb{R}^n \setminus B_{2k}(0) \text{ and } |D\phi_k| \leq c/k. \tag{3.1}$$

Then

$$\lim_{k\to\infty} \|(\phi_k u) - u\|_{BV_{\frac{n}{n-1}}(\mathbb{R}^n)} = 0, \tag{3.2}$$

and for each fixed $k > 0$ we have

$$\lim_{j\to\infty} \|(\phi_k u) \wedge j - \phi_k u\|_{BV_{\frac{n}{n-1}}(\mathbb{R}^n)} = 0. \tag{3.3}$$

In particular, $BV_c^\infty(\mathbb{R}^n)$ is dense in $BV_{\frac{n}{n-1}}(\mathbb{R}^n)$.

With this density result we can show that:

$$BV(\mathbb{R}^n)^* \text{ and } BV_{\frac{n}{n-1}}(\mathbb{R}^n)^* \text{ are isometrically isomorphic.} \tag{3.4}$$

Indeed, we define the map

$$S : BV_{\frac{n}{n-1}}(\mathbb{R}^n)^* \to BV(\mathbb{R}^n)^*$$

as

$$S(T) = T \llcorner BV(\mathbb{R}^n).$$

Clearly (see [**PT17**, Corollary 3.3]), Theorem 3.1 implies that S is an isometry.

We now proceed to make precise our definitions of measures in $\dot{W}^{1,1}(\mathbb{R}^n)^*$ and $BV_{\frac{n}{n-1}}(\mathbb{R}^n)^*$.

DEFINITION 3.2. We let

$$\begin{aligned}\mathcal{M}_{loc}(\mathbb{R}^n) \cap \dot{W}^{1,1}(\mathbb{R}^n)^* := \{T \in \dot{W}^{1,1}(\mathbb{R}^n)^* : T(\varphi) \\ = \int_{\mathbb{R}^n} \varphi d\mu \text{ for some } \mu \in \mathcal{M}_{loc}(\mathbb{R}^n), \forall \varphi \in C_c^\infty(\mathbb{R}^n)\}.\end{aligned}$$

Therefore, if $\mu \in \mathcal{M}_{loc}(\mathbb{R}^n) \cap \dot{W}^{1,1}(\mathbb{R}^n)^*$, then the action $\langle \mu, u \rangle$ can be uniquely defined for all $u \in \dot{W}^{1,1}(\mathbb{R}^n)$.

DEFINITION 3.3. We let

$$\begin{aligned}\mathcal{M}_{loc} \cap BV_{\frac{n}{n-1}}(\mathbb{R}^n)^* := \{T \in BV_{\frac{n}{n-1}}(\mathbb{R}^n)^* : T(\varphi) \\ = \int_{\mathbb{R}^n} \varphi^* d\mu \text{ for some } \mu \in \mathcal{M}_{loc}(\mathbb{R}^n), \forall \varphi \in BV_c^\infty(\mathbb{R}^n)\},\end{aligned}$$

where φ^* is the precise representative of φ in $BV_c^\infty(\mathbb{R}^n)$ (see **[AFP**, Corollary 3.8] for the definition of precise representative). Thus, if $\mu \in \mathcal{M}_{loc}(\mathbb{R}^n) \cap BV_{\frac{n}{n-1}}(\mathbb{R}^n)^*$, then the action $\langle \mu, u \rangle$ can be uniquely defined for all $u \in BV_{\frac{n}{n-1}}(\mathbb{R}^n)$.

In Definition 3.3, if we use $C_c^\infty(\mathbb{R}^n)$ instead of $BV_c^\infty(\mathbb{R}^n)$, then by the Hahn-Banach Theorem there exists a non-zero $T \in BV_{\frac{n}{n-1}}(\mathbb{R}^n)^*$ that is represented by the zero measure, which would cause a problem of injectivity in Theorem 3.6.

The following lemma (see **[PT17**, Lemma 4.1]) characterizes all the distributions in $\dot{W}^{1,1}(\mathbb{R}^n)^*$ and is a key ingredient in the characterization of the measures in $BV_{\frac{n}{n-1}}(\mathbb{R}^n)^*$. We recall that $\dot{W}^{1,1}(\mathbb{R}^n)$ is the homogeneous Sobolev space introduced in Definition 2.4 and that it is also known as the G space in image processing.

LEMMA 3.4. *The distribution T belongs to $\dot{W}^{1,1}(\mathbb{R}^n)^*$ if and only if $T = div\, \boldsymbol{F}$ for some vector field $\boldsymbol{F} \in L^\infty(\mathbb{R}^n, \mathbb{R}^n)$. Moreover,*

$$\|T\|_{\dot{W}^{1,1}(\mathbb{R}^n)^*} = \min\{\|\boldsymbol{F}\|_{L^\infty(\mathbb{R}^n,\mathbb{R}^n)}\},$$

where the minimum is taken over all $\boldsymbol{F} \in L^\infty(\mathbb{R}^n, \mathbb{R}^n)$ such that $\operatorname{div} \boldsymbol{F} = T$. Here we use the norm

$$\|\boldsymbol{F}\|_{L^\infty(\mathbb{R}^n,\mathbb{R}^n)} := \left\|(F_1^2 + F_2^2 + \cdots + F_n^2)^{1/2}\right\|_{L^\infty(\mathbb{R}^n)} \quad \textit{for } \boldsymbol{F} = (F_1, \ldots, F_n).$$

The following theorem characterizes all the signed measures in $BV_{\frac{n}{n-1}}(\mathbb{R}^n)^*$ (see **[PT17**, Theorem 4.4]. This result was first proven in Phuc-Torres **[PT]** for the space $BV(\mathbb{R}^n)^*$ with no sharp control on the involving constants. In Phuc-Torres **[PT17]** we offered a new and direct proof of **(i)** $\Rightarrow$ **(ii)**. We also clarified the first part of **(iii)**. The second part of **(iii)** relies on a delicate argument using the coarea formula. Moreover, our proof of **(ii)** $\Rightarrow$ **(iii)** yields a sharp constant that will be needed for the proof of Theorem 3.6 below. The density obtained in Theorem 3.1 allows to show **(iii)** $\Rightarrow$ **(iv)**. Indeed, we note that the main idea in the proof of Theorem 3.5 is to show first that, if $div\, \boldsymbol{F} = \mu$, then the action of the measure μ can be defined for any function $u \in BV_c^\infty(\mathbb{R}^n)$ and thus, since $BV_c^\infty(\mathbb{R}^n)$ is dense in $BV_{\frac{n}{n-1}}(\mathbb{R}^n)$, the action of μ can be extended to all $BV_{\frac{n}{n-1}}(\mathbb{R}^n)$, which shows that $\mu \in BV_{\frac{n}{n-1}}(\mathbb{R}^n)^*$. Finally, Lemma 3.4 yields **(iv)** $\Rightarrow$ **(i)**.

THEOREM 3.5. *Let $\mu \in \mathcal{M}_{loc}(\mathbb{R}^n)$ be a locally finite signed measure. The following are equivalent:*

(i) *There exists a vector field $\boldsymbol{F} \in L^\infty(\mathbb{R}^n, \mathbb{R}^n)$ such that $\operatorname{div} \boldsymbol{F} = \mu$ in the sense of distributions.*

(ii) *There is a constant C such that*

$$|\mu(U)| \leq C\, \mathcal{H}^{n-1}(\partial U)$$

for any smooth bounded open (or closed) set U with $\mathcal{H}^{n-1}(\partial U) < +\infty$.

(iii) $\mathcal{H}^{n-1}(A) = 0$ *implies* $\|\mu\|(A) = 0$ *for all Borel sets A and there is a constant C such that, for all $u \in BV_c^\infty(\mathbb{R}^n)$,*

$$|\langle \mu, u\rangle| := \left|\int_{\mathbb{R}^n} u^* d\mu\right| \le C \int_{\mathbb{R}^n} |Du|,$$

where u^* is the precise representative in the class of u that is defined $\mathcal{H}^{n-1}$-almost everywhere.

(iv) $\mu \in BV_{\frac{n}{n-1}}(\mathbb{R}^n)^*$. *The action of μ on any $u \in BV_{\frac{n}{n-1}}(\mathbb{R}^n)$ is defined (uniquely) as*

$$\langle \mu, u\rangle := \lim_{k\to\infty} \langle \mu, u_k\rangle = \lim_{k\to\infty} \int_{\mathbb{R}^n} u_k^* d\mu,$$

where $u_k \in BV_c^\infty(\mathbb{R}^n)$ converges to u in $BV_{\frac{n}{n-1}}(\mathbb{R}^n)$. In particular, if $u \in BV_c^\infty(\mathbb{R}^n)$ then

$$\langle \mu, u\rangle = \int_{\mathbb{R}^n} u^* d\mu,$$

and moreover, if μ is a non-negative measure then, for all $u \in BV_{\frac{n}{n-1}}(\mathbb{R}^n)$,

$$\langle \mu, u\rangle = \int_{\mathbb{R}^n} u^* d\mu.$$

We recall the spaces introduced in Definitions 3.2 and 3.3. With the previous theorem, we can prove the following new result (see [**PT17**, Theorem 4.7]).

THEOREM 3.6. *Let $\mathcal{E} := \mathcal{M}_{loc}(\mathbb{R}^n) \cap BV_{\frac{n}{n-1}}(\mathbb{R}^n)^*$ and $\mathcal{F} := \mathcal{M}_{loc}(\mathbb{R}^n) \cap \dot{W}^{1,1}(\mathbb{R}^n)^*$. Then $\mathcal{E}$ and $\mathcal{F}$ are isometrically isomorphic.*

In the following remark we discuss the connection between Theorem 3.5, Theorem 3.6 and image processing.

REMARK 3.7. The space $\dot{W}^{1,1}(\mathbb{R}^n)^*$ is denoted as the G space in image processing (see Meyer [**ME**]), and it plays a key role in modeling the noise of an image. It is mentioned in [**ME**] that it is more convenient to work with G instead of $BV_{\frac{n}{n-1}}(\mathbb{R}^n)^*$. Indeed, except for the characterization of the (signed) measures treated in this paper and the results in De Pauw-Torres [**DP**], the full characterization of $BV_{\frac{n}{n-1}}(\mathbb{R}^n)^*$ is unknown. However, G can be easily characterized; see Lemma 3.4. Our previous results Theorem 3.5 and Theorem 3.6 show that, when restricted to measures, both spaces coincide. Moreover, the norm of any (signed) measure $\mu \in G$ can be computed as

$$\|\mu\|_G = \sup \frac{|\mu(U)|}{\mathcal{H}^{n-1}(\partial U)}, \tag{3.5}$$

where the sup is taken over all bounded open sets $U \subset \mathbb{R}^n$ with smooth boundary and $\mathcal{H}^{n-1}(\partial U) < +\infty$. Hence, our results give an alternative to the more abstract computation of $\|\mu\|_G$ given, by Lemma 3.4, as

$$\|\mu\|_G = \min\{\|\boldsymbol{F}\|_{L^\infty(\mathbb{R}^n,\mathbb{R}^n)}\},$$

where the minimum is taken over all $\boldsymbol{F} \in L^\infty(\mathbb{R}^n, \mathbb{R}^n)$ such that $div\boldsymbol{F} = T$. We refer the reader to Kindermann-Osher-Xu [**KOX**] for an algorithm based on the level set method to compute (3.5) for the case when μ is a function $f \in L^2(\mathbb{R}^2)$ with zero mean. Also, in the two-dimensional case, when μ is a function $f \in L^2(\mathbb{R}^2)$, the isometry of measures in Theorem 3.6 could be deduced from [**HM**, Lemma 3.1].

4. On an issue raised by Meyers and Ziemer

Using Theorem 3.5, we constructed in **[PT17**, Section 5] a locally integrable function f such that $f \in BV(\mathbb{R}^n)^*$ but $|f| \notin BV(\mathbb{R}^n)^*$. We mention that this kind of highly oscillatory function appeared in [**MV**] in a different context. It is clear that if μ is a locally finite signed Radon measure then $\|\mu\| \in BV(\mathbb{R}^n)^*$ implies that $\mu \in BV(\mathbb{R}^n)^*$. The converse was unknown to Meyers and Ziemer as they raised this issue in their classical paper [**MZ**, page 1356]. The following Proposition shows that the converse does not hold true in general.

PROPOSITION 4.1. *Let* $f(x) = \epsilon|x|^{-1-\epsilon}\sin(|x|^{-\epsilon})+(n-1)|x|^{-1}\cos(|x|^{-\epsilon})$, *where* $0 < \epsilon < n-1$ *is fixed. Then*

$$f(x) = \operatorname{div}\,[x|x|^{-1}\cos(|x|^{-\epsilon})]. \tag{4.1}$$

Moreover, there exists a sequence $\{r_k\}$ *decreasing to zero such that*

$$\int_{B(0,r_k)} f^+(x)dx \geq c\, r_k^{n-1-\epsilon} \tag{4.2}$$

for a constant $c = c(n,\epsilon) > 0$ *independent of* k. *Here* f^+ *is the positive part of* f. *Thus by Theorem 3.5 we see that* f *belongs to* $BV(\mathbb{R}^n)^*$, *whereas* $|f|$ *does not.*

5. Characterization of a subspace of $BV_{\frac{n}{n-1}}(\mathbb{R}^n)^*$ whose elements are the divergence of continuous vector fields vanishing at infinity

In this section we characterize a subspace of $BV_{\frac{n}{n-1}}(\mathbb{R}^n)^*$, denoted as $CH_0(\mathbb{R}^n)$, and consisting of all *charges vanishing at infinity.* We start by introducing a notion of convergence in $BV_{\frac{n}{n-1}}(\mathbb{R}^n)$ and the definition of *charge vanishing at infinity.*

DEFINITION 5.1. Given a sequence $\{u_j\}$ in $BV_{\frac{n}{n-1}}(\mathbb{R}^n)$, we write $u_j \twoheadrightarrow 0$ whenever

(1) $\sup_j \|Du_j\|(\mathbb{R}^n) < \infty$;
(2) $u_j \rightharpoonup 0$ weakly in $L^{\frac{n}{n-1}}(\mathbb{R}^n)$.

A *charge vanishing at infinity* is a linear functional $T : BV_{\frac{n}{n-1}}(\mathbb{R}^n) \to \mathbb{R}$ such that $\langle u_j, T\rangle \to 0$ whenever $u_j \twoheadrightarrow 0$. The set of charges vanishing at infinity is denoted as $CH_0(\mathbb{R}^n)$.

We showed in [**DPT**] that the equation $div\, \boldsymbol{F} = T$ has a solution $\boldsymbol{F} \in C_0(\mathbb{R}^n, \mathbb{R}^n)$ if and only if $T \in CH_0(\mathbb{R}^n)$, where $C_0(\mathbb{R}^n, \mathbb{R}^n)$ is the space of continuous vector fields vanishing at infinity. We recall that $\boldsymbol{F} \in C_0(\mathbb{R}^n, \mathbb{R}^n)$ if and only if, for every $\varepsilon > 0$, there exists a compact set $K \subset \mathbb{R}^n$ such that $|\boldsymbol{F}(x)| \leq \varepsilon$ whenever $x \in \mathbb{R}^n \setminus K$. The terminology *charge vanishing at infinity* was motivated by the fact that, given $T \in CH_0(\mathbb{R}^n)$ and $\varepsilon > 0$, there exists a compact set $K \subset \mathbb{R}^n$ such that

$$|\langle u, T\rangle| \leq \varepsilon \,\|Du\|\,(\mathbb{R}^n),$$

whenever $u \in BV_{\frac{n}{n-1}}(\mathbb{R}^n)$ and $K \cap \operatorname{supp} u = \emptyset$.

A compactness property can be proven in $BV_{\frac{n}{n-1}}(\mathbb{R}^n)$ (see [**DPT**, Proposition 2.6]) which states that if $\{u_j\}$ is a bounded sequence in $BV_{\frac{n}{n-1}}(\mathbb{R}^n)$, i.e. $\sup_j \|Du_j\|\,(\mathbb{R}^n) < \infty$, then there exists a subsequence $\{u_{j_k}\}$ of $\{u_j\}$ and $u \in BV_{\frac{n}{n-1}}(\mathbb{R}^n)$ such that $u_{j_k} \twoheadrightarrow u$. The compactness in $BV_{\frac{n}{n-1}}(\mathbb{R}^n)$ implies that

$$\|T\|_{CH_0} := \sup\{\langle u, T\rangle : u \in BV_{\frac{n}{n-1}}(\mathbb{R}^n) \text{ and } \|Du\|(\mathbb{R}^n) \leq 1\} < \infty \tag{5.1}$$

and hence $\|\cdot\|_{CH_0}$ defines a norm in $CH_0(\mathbb{R}^n)$. Moreover, $CH_0(\mathbb{R}^n)$ with the norm $\|\cdot\|_{CH_0}$ is a Banach space. From (5.1) we see that

$$CH_0(\mathbb{R}^n) \subset BV_{\frac{n}{n-1}}(\mathbb{R}^n)^*.$$

We now consider the divergence operator

$$div : C_0(\mathbb{R}^n, \mathbb{R}^n) \to CH_0(\mathbb{R}^n)$$

given as

$$div(\boldsymbol{F}) : BV_{\frac{n}{n-1}}(\mathbb{R}^n) \to \mathbb{R}, \quad \boldsymbol{F} \in C_0(\mathbb{R}^n, \mathbb{R}^n),$$

$$div(\boldsymbol{F})(u) = -\int_{\mathbb{R}^n} \langle \boldsymbol{F}, d(Du)\rangle, \ u \in BV_{\frac{n}{n-1}}(\mathbb{R}^n).$$

It can be proven that this operator is well defined. Moreover, we have that

$$div : C_0(\mathbb{R}^n, \mathbb{R}^n) \to CH_0(\mathbb{R}^n)$$

is a bounded linear operator satisfying

$$\|div(\boldsymbol{F})\|_{CH_0} \leq \|\boldsymbol{F}\|_\infty .$$

The question now is whether the divergence operator defined above is surjective. Indeed, this is true so we have:

THEOREM 5.2. *There exists* $\boldsymbol{F} \in C_0(\mathbb{R}^n, \mathbb{R}^n)$ *such that*

$$div\, \boldsymbol{F} = T$$

if and only if $T \in CH_0(\mathbb{R}^n)$.

Since

$$CH_0(\mathbb{R}^n) \subset BV_{\frac{n}{n-1}}(\mathbb{R}^n)^*,$$

we have then characterized a closed subspace of $BV_{\frac{n}{n-1}}(\mathbb{R}^n)^*$. Moreover, since (see [**DPT**, Proposition 3.4]):

$$L^n(\mathbb{R}^n) \subset CH_0(\mathbb{R}^n),$$

the theorem implies that to each $f \in L^n(\mathbb{R}^n)$ there corresponds a continuous vector field vanishing at infinity, $\boldsymbol{F} \in C_0(\mathbb{R}^n, \mathbb{R}^n)$, such that

$$div\, \boldsymbol{F} = f \tag{5.2}$$

in the sense that

$$-\int_{\mathbb{R}^n} \langle \boldsymbol{F}, d(Du)\rangle = \int_{\mathbb{R}^n} fu, \quad u \in BV_{\frac{n}{n-1}}(\mathbb{R}^n), \tag{5.3}$$

and in particular in the sense of distributions. Therefore, (5.2) addresses the solvability of the critical case $p = n$ discussed in the introduction. We now present the main idea of the proof of Theorem 5.2 (see [**DPT**, Theorem 6.1] for the details), which uses the adjoint operator:

$$\begin{array}{ccc} CH_0(\mathbb{R}^n)^* & \xrightarrow{div^*} & C_0(\mathbb{R}^n;\mathbb{R}^n)^* \\ ev \Big\| & & \Big\| \\ BV_{\frac{n}{n-1}}(\mathbb{R}^n) & \xrightarrow[-D]{} & \mathcal{M}(\mathbb{R}^n, \mathbb{R}^n) \end{array}$$

and the fact that the evaluation map

$$ev : BV_{\frac{n}{n-1}}(\mathbb{R}^n) \to CH_0(\mathbb{R}^n)^*$$

is a bijection, where

$$\langle T, ev(u)\rangle = \langle u, T\rangle, \quad u \in BV_{\frac{n}{n-1}}(\mathbb{R}^n), \quad T \in CH_0(\mathbb{R}^n).$$

Moreover, the evaluation map is in fact an isomorphism of the Banach spaces $BV_{\frac{n}{n-1}}(\mathbb{R}^n)$ and $CH_0(\mathbb{R}^n)^*$, according to the Open Mapping Theorem. Since the Range of div is dense in $CH_0(\mathbb{R}^n)$, in order to show that $div\,(C_0(\mathbb{R}^n, \mathbb{R}^n))$ is closed in $CH_0(\mathbb{R}^n)$, and hence div is a surjective operator, it suffices to show that the Range of the adjoint operator div^* is closed in $C_0(\mathbb{R}^n, \mathbb{R}^n)^*$, according to the Closed Range Theorem. This can be seen as follows. Let $\{\alpha_j\}$ be a sequence in $CH_0(\mathbb{R}^n)^*$ such that $div^*(\alpha_j) \to \mu$. Let $\{u_j\}$ in $BV_{\frac{n}{n-1}}(\mathbb{R}^n)$ such that $\alpha_j = ev(u_j)$. Hence

$$\|div^*(\alpha_j)\|_{\mathcal{M}} = \|(div^* \circ ev)(u_j)\|_{\mathcal{M}} = \|Du_j\|_{\mathcal{M}}\,.$$

Since $\{div^*(\alpha_j)\}$ is bounded then $\sup_j \|Du_j\|_{\mathcal{M}} < \infty$. Therefore, the compactness property described at the beginning of this section implies that there exists a subsequence $\{u_{j_k}\}$ and $u \in BV_{\frac{n}{n-1}}(\mathbb{R}^n)$ such that $u - u_{j_k} \twoheadrightarrow 0$. Since

$$\int_{\mathbb{R}^n} \langle \boldsymbol{F}, d(Du)\rangle = -\int_{\mathbb{R}^n} u\,div\,\boldsymbol{F} = -\lim_k \int_{\mathbb{R}^n} u_{j_k}\,div\,\boldsymbol{F} = \lim_k \int_{\mathbb{R}^n} \langle \boldsymbol{F}, d(Du_{j_k})\rangle\,,$$

then $\int_{\mathbb{R}^n} \langle \boldsymbol{F}, d(Du)\rangle = \int_{\mathbb{R}^n} \langle \boldsymbol{F}, d\mu\rangle$ since $Du_j \to \mu$, from which it follows that $Du = \mu$.

6. Characterization of (signed) measures in $BV_0(\Omega)^*$

In this section we let $\Omega \subset \mathbb{R}^n$ be a bounded open set with Lipschitz boundary. We recall that functions in $BV(\Omega)$ (see for example [**G**, Theorem 2.10] and [**ABM**, Theorem 10.2.1]) have traces on $\partial\Omega$. More precisely, if $u \in BV(\Omega)$, there exists a function $\varphi \in L^1(\partial\Omega)$ such that, for $\mathcal{H}^{n-1}$-almost every $x \in \partial\Omega$,

$$\lim_{r\to 0} r^{-n} \int_{B(x,r)\cap\Omega} |u(y) - \varphi(x)|dy = 0.$$

From the construction of the trace φ (see [**G**, Lemma 2.4]), we see that φ is uniquely determined. Therefore, we have a well defined operator

$$\gamma_0 : BV(\Omega) \to L^1(\partial\Omega).$$

The trace operator γ_0 is continuous from $BV(\Omega)$ equipped with the intermediate convergence onto $L^1(\Omega)$ equipped with the strong convergence. By the intermediate convergence of $BV(\Omega)$ we mean that the sequence $\{u_k\} \in BV(\Omega)$ converges to $u \in BV(\Omega)$ in the sense of intermediate (or strict) convergence if

$$u_k \to u \text{ strongly in } L^1(\Omega) \text{ and } \int_\Omega |Du_k| \to \int_\Omega |Du|.$$

It is natural to define $BV_0(\Omega)$ as follows:

DEFINITION 6.1. Let

$$BV_0(\Omega) = \ker(\gamma_0).$$

We also define another BV function space with a zero boundary condition.

DEFINITION 6.2. Let

$$\mathbb{BV}_0(\Omega) := \overline{C_c^\infty(\Omega)},$$

where the closure is taken with respect to the intermediate convergence of $BV(\Omega)$.

Both spaces $\mathbb{BV}_0(\Omega)$ and $BV_0(\Omega)$ are the same. Clearly, $\mathbb{BV}_0(\Omega) \subset BV_0(\Omega)$ holds by the continuity of the trace operator. In order to prove $BV_0(\Omega) \subset \mathbb{BV}_0(\Omega)$, it is necessary to show that the space $BV_c(\Omega)$ is dense in $BV_0(\Omega)$ in the strong topology of $BV(\Omega)$. This can be proven by using Giusti [**G**, Inequality (2.10)] (see [**PT17**, Theorem 6.3] for details). With this result at hand, given $u \in BV_0(\Omega)$, we can find by convolution a sequence in $C_c^\infty(\Omega)$ that converges to u in the intermediate convergence. Hence, we have

$$\mathbb{BV}_0(\Omega) = BV_0(\Omega). \tag{6.1}$$

We note that (6.1) implies the following Sobolev's inequality for functions in $BV_0(\Omega)$

$$\|u\|_{L^{\frac{n}{n-1}}(\Omega)} \leq C \|Du\|(\Omega), \ u \in BV_0(\Omega), \tag{6.2}$$

for a constant $C = C(n)$.

Also, from (6.2), we see that the full norm $\|u\|_{BV(\Omega)} = \|u\|_{L^1(\Omega)} + \|Du\|(\Omega)$ is equivalent to $\|Du\|(\Omega)$ whenever $u \in BV_0(\Omega)$ (or $\mathbb{BV}_0(\Omega)$) and Ω is a bounded open set with Lipschitz boundary. Thus, for the rest of the paper we will equip $BV_0(\Omega)$ with the homogeneous norm:

$$\|u\|_{BV_0(\Omega)} = \|Du\|(\Omega). \tag{6.3}$$

The fact that $BV_c(\Omega)$ is dense in $BV_0(\Omega)$ in the strong topology of $BV(\Omega)$ implies the following Corollary (see [**PT17**, Lemma 6.5]) which is the analogous (for bounded domains) to Theorem 3.1, and it is a key element in the characterization of measures in $BV_0(\Omega)^*$.

COROLLARY 6.3. *Let Ω be any bounded open set with Lipschitz boundary. Then $BV_c^\infty(\Omega)$ is dense in $BV_0(\Omega)$.*

We now proceed to make precise the definitions of measures in the spaces $W_0^{1,1}(\Omega)^*$ and $BV_0(\Omega)^*$.

DEFINITION 6.4. For a bounded open set Ω with Lipschitz boundary, we let

$$\begin{aligned}\mathcal{M}_{loc}(\Omega) \cap W_0^{1,1}(\Omega)^* := \{T \in W_0^{1,1}(\Omega)^* : T(\varphi) \\ = \int_\Omega \varphi d\mu \text{ for some } \mu \in \mathcal{M}_{loc}(\Omega), \forall \varphi \in C_c^\infty(\Omega)\}.\end{aligned}$$

Therefore, if $\mu \in \mathcal{M}_{loc}(\Omega) \cap W_0^{1,1}(\Omega)^*$, then the action $\langle \mu, u \rangle$ can be uniquely defined for all $u \in W_0^{1,1}(\Omega)$.

DEFINITION 6.5. For a bounded open set Ω with Lipschitz boundary, we let

$$\begin{aligned}\mathcal{M}_{loc}(\Omega) \cap BV_0(\Omega)^* := \{T \in BV_0(\Omega)^* : T(\varphi) \\ = \int_\Omega \varphi^* d\mu \text{ for some } \mu \in \mathcal{M}_{loc}(\Omega), \forall \varphi \in BV_c^\infty(\Omega)\},\end{aligned}$$

where φ^* is the precise representative of φ. Thus, if $\mu \in \mathcal{M}_{loc}(\Omega) \cap BV_0(\Omega)^*$, then the action $\langle \mu, u \rangle$ can be uniquely defined for all $u \in BV_0(\Omega)$.

We will use the following characterization of $W_0^{1,1}(\Omega)^*$ whose proof is analogous to that of Lemma 3.4.

LEMMA 6.6. *Let Ω be any bounded open set with Lipschitz boundary. The distribution T belongs to $W_0^{1,1}(\Omega)^*$ if and only if $T = \operatorname{div} \boldsymbol{F}$ for some vector field $\boldsymbol{F} \in L^\infty(\Omega, \mathbb{R}^n)$. Moreover,*

$$\|T\|_{W_0^{1,1}(\Omega)^*} = \min\{\|\boldsymbol{F}\|_{L^\infty(\Omega,\mathbb{R}^n)}\},$$

where the minimum is taken over all $\boldsymbol{F} \in L^\infty(\Omega, \mathbb{R}^n)$ such that $\operatorname{div} \boldsymbol{F} = T$. Here we use the norm

$$\|\boldsymbol{F}\|_{L^\infty(\Omega,\mathbb{R}^n)} := \left\|(F_1^2 + F_2^2 + \cdots + F_n^2)^{1/2}\right\|_{L^\infty(\Omega)} \quad \text{for } \boldsymbol{F} = (F_1, \ldots, F_n).$$

We are now ready to state the main result of this section (see [**PT17**, Theorem 7.4]), whose proof uses Corollary 6.3 (for the proof of (**iii**) $\Rightarrow$ (**iv**)) and Lemma 6.6 (for the proof of (**iv**) $\Rightarrow$ (**i**)).

THEOREM 6.7. *Let Ω be any bounded open set with Lipschitz boundary and $\mu \in \mathcal{M}_{loc}(\Omega)$. Then, the following are equivalent:*

(i) *There exists a vector field $\boldsymbol{F} \in L^\infty(\Omega, \mathbb{R}^n)$ such that $\operatorname{div} \boldsymbol{F} = \mu$.*

(ii) *$|\mu(U)| \leq C\mathcal{H}^{n-1}(\partial U)$ for any smooth open (or closed) set $U \subset\subset \Omega$ with $\mathcal{H}^{n-1}(\partial U) < +\infty$.*

(iii) *$\mathcal{H}^{n-1}(A) = 0$ implies $\|\mu\|(A) = 0$ for all Borel sets $A \subset \Omega$ and there is a constant C such that, for all $u \in BV_c^\infty(\Omega)$,*

$$|\langle \mu, u\rangle| := \left|\int_\Omega u^* d\mu\right| \leq C \int_\Omega |Du|,$$

where u^* is the precise representative in the class of u that is defined $\mathcal{H}^{n-1}$-almost everywhere.

(iv) *$\mu \in BV_0(\Omega)^*$. The action of μ on any $u \in BV_0(\Omega)$ is defined (uniquely) as*

$$\langle \mu, u\rangle := \lim_{k\to\infty} \langle \mu, u_k\rangle = \lim_{k\to\infty} \int_\Omega u_k^* d\mu,$$

where $u_k \in BV_c^\infty(\Omega)$ converges to u in $BV_0(\Omega)$. In particular, if $u \in BV_c^\infty(\Omega)$ then

$$\langle \mu, u\rangle = \int_\Omega u^* d\mu,$$

and moreover, if μ is a non-negative measure then, for all $u \in BV_0(\Omega)$,

$$\langle \mu, u\rangle = \int_\Omega u^* d\mu.$$

REMARK 6.8. If Ω is a bounded domain containing the origin then the function f given in Proposition 4.1 belongs to $BV_0(\Omega)^*$ but $|f|$ does not.

Theorem 6.7 immediately imply the following new result which states that the set of measures in $BV_0(\Omega)^*$ coincides with that of $W_0^{1,1}(\Omega)^*$.

THEOREM 6.9. *The normed spaces $\mathcal{M}_{loc}(\Omega) \cap BV_0(\Omega)^*$ and $\mathcal{M}_{loc}(\Omega) \cap W_0^{1,1}(\Omega)^*$ are isometrically isomorphic.*

The proof of Theorem 6.9 is similar to that of Theorem 3.6 but this time one uses Theorem 6.7 and Corollary 6.3 in place of Theorem 3.5 and Theorem 3.1, respectively (see [**PT17**, Theorem 7.6] for the details of the proof).

7. Finite measures in $BV(\Omega)^*$

The characterization of measures in the space $BV(\Omega)^*$, where $BV(\Omega)$ is the space of functions of bounded variation in a bounded open set Ω with Lipschitz boundary (but without the condition of having zero trace on $\partial\Omega$), offers more difficulties since we have to work with the full norm $\|u\|_{BV(\Omega)} = \|u\|_{L^1(\Omega)} + \|Du\|(\Omega)$. However, we can still characterize all *finite* signed measures that belong to $BV(\Omega)^*$. Note that the finiteness condition here is necessary at least for *positive* measures in $BV(\Omega)^*$. By a measure $\mu \in BV(\Omega)^*$ we mean that the inequality

$$\left| \int_\Omega u^* d\mu \right| \leq C \|u\|_{BV(\Omega)}$$

holds for all $u \in BV^\infty(\Omega)$. Since $BV^\infty(\Omega)$ is dense in $BV(\Omega)$ in the strong topology of $BV(\Omega)$, we see that such a μ can be uniquely extended to be a continuous linear functional in $BV(\Omega)$. We have the following (see [**PT17**, Theorem 8.2]):

THEOREM 7.1. *Let Ω be an open set with Lipschitz boundary and let μ be a finite signed measure in Ω. Extend μ by zero to $\mathbb{R}^n \setminus \Omega$ by setting $\|\mu\|(\mathbb{R}^n \setminus \Omega) = 0$. Then, $\mu \in BV(\Omega)^*$ if and only if*

$$|\mu(U)| \leq C\mathcal{H}^{n-1}(\partial U) \tag{7.1}$$

for every smooth open set $U \subset \mathbb{R}^n$ and a constant $C = C(\Omega, \mu)$.

The main ingredient in the proof of Theorem 7.1 is the following result, concerning the extension of BV functions, and whose proof can be found in [**Z**, Lemma 5.10.14]:

LEMMA 7.2. *Let Ω be an open set with Lipschitz boundary and $u \in BV(\Omega)$. Then, the extension of u to $\mathbb{R}^n$ defined by*

$$u_0(x) = \begin{cases} u(x), & x \in \Omega \\ 0, & x \in \mathbb{R}^n \setminus \Omega \end{cases}$$

satisfies that $u_0 \in BV(\mathbb{R}^n)$ and

$$\|u_0\|_{BV(\mathbb{R}^n)} \leq C \|u\|_{BV(\Omega)},$$

where $C = C(\Omega)$.

It is easy to see that if μ is a positive measure in $BV(\Omega)^*$ then its action on $BV(\Omega)$ is given by

$$\langle \mu, u \rangle = \int_\Omega u^* d\mu, \text{ for all } u \in BV(\Omega).$$

References

[ACV08] L. Ambrosio, A. Colesanti, and E. Villa, *Outer Minkowski content for some classes of closed sets*, Math. Ann. **342** (2008), no. 4, 727–748. MR2443761

[AFP] L. Ambrosio, N. Fusco, and D. Pallara, *Functions of bounded variation and free discontinuity problems*, Oxford Mathematical Monographs, The Clarendon Press, Oxford University Press, New York, 2000. MR1857292

[ABM] H. Attouch, G. Buttazzo, and G. Michaille, *Variational analysis in Sobolev and BV spaces*, MPS/SIAM Series on Optimization, vol. 6, Society for Industrial and Applied Mathematics (SIAM), Philadelphia, PA; Mathematical Programming Society (MPS), Philadelphia, PA, 2006. Applications to PDEs and optimization. MR2192832

[BB] J. Bourgain and H. Brezis, *On the equation* $\operatorname{div} Y = f$ *and application to control of phases*, J. Amer. Math. Soc. **16** (2003), no. 2, 393–426. MR1949165

[BMR] P. Bousquet, P. Mironescu, and E. Russ, *A limiting case for the divergence equation*, Math. Z. **274** (2013), no. 1-2, 427–460. MR3054337

[CTZ] G.-Q. Chen, M. Torres, and W. P. Ziemer, *Gauss-Green theorem for weakly differentiable vector fields, sets of finite perimeter, and balance laws*, Comm. Pure Appl. Math. **62** (2009), no. 2, 242–304. MR2468610

[DFT] B. Dacorogna, N. Fusco, and L. Tartar, *On the solvability of the equation* $\operatorname{div} u = f$ *in* L^1 *and in* C^0, Atti Accad. Naz. Lincei Cl. Sci. Fis. Mat. Natur. Rend. Lincei (9) Mat. Appl. **14** (2003), no. 3, 239–245 (2004). Renato Caccioppoli and modern analysis. MR2064270

[D] J. Dieudonné, *Treatise on analysis. Vol. II*, Translated from the French by I. G. Macdonald. Pure and Applied Mathematics, Vol. 10-II, Academic Press, New York-London, 1970. MR0258551

[DU] J. Diestel and J. J. Uhl Jr., *Vector measures*, American Mathematical Society, Providence, R.I., 1977. With a foreword by B. J. Pettis; Mathematical Surveys, No. 15. MR0453964

[DP] T. De Pauw, *On SBV dual*, Indiana Univ. Math. J. **47** (1998), no. 1, 99–121. MR1631541

[DPP] T. De Pauw and W. F. Pfeffer, *Distributions for which* $\operatorname{div} v = F$ *has a continuous solution*, Comm. Pure Appl. Math. **61** (2008), no. 2, 230–260. MR2368375

[DPT] T. De Pauw and M. Torres, *On the distributional divergence of vector fields vanishing at infinity*, Proc. Roy. Soc. Edinburgh Sect. A **141** (2011), no. 1, 65–76. MR2773439

[EG] L. C. Evans and R. F. Gariepy, *Measure theory and fine properties of functions*, Studies in Advanced Mathematics, CRC Press, Boca Raton, FL, 1992. MR1158660

[FS] N. Fusco and D. Spector, *A remark on an integral characterization of the dual of* BV, J. Math. Anal. Appl. **457** (2018), no. 2, 1370–1375. MR3705358

[G] E. Giusti, *Minimal surfaces and functions of bounded variation*, Monographs in Mathematics, vol. 80, Birkhäuser Verlag, Basel, 1984. MR775682

[HM] A. Haddad and Y. Meyer, *An improvement of Rudin-Osher-Fatemi model*, Appl. Comput. Harmon. Anal. **22** (2007), no. 3, 319–334. MR2311857

[KOX] S. Kindermann, S. Osher, and J. Xu, *Denoising by BV-duality*, J. Sci. Comput. **28** (2006), no. 2-3, 411–444. MR2272638

[MV] V. G. Maz'ya and I. E. Verbitsky, *The Schrödinger operator on the energy space: boundedness and compactness criteria*, Acta Math. **188** (2002), no. 2, 263–302. MR1947894

[M] C. T. McMullen, *Lipschitz maps and nets in Euclidean space*, Geom. Funct. Anal. **8** (1998), no. 2, 304–314. MR1616159

[ME] Y. Meyer, *Oscillating patterns in image processing and nonlinear evolution equations*, University Lecture Series, vol. 22, American Mathematical Society, Providence, RI, 2001. The fifteenth Dean Jacqueline B. Lewis memorial lectures. MR1852741

[MZ] N. G. Meyers and W. P. Ziemer, *Integral inequalities of Poincaré and Wirtinger type for* BV *functions*, Amer. J. Math. **99** (1977), no. 6, 1345–1360. MR0507433

[PT] N. C. Phuc and M. Torres, *Characterizations of the existence and removable singularities of divergence-measure vector fields*, Indiana Univ. Math. J. **57** (2008), no. 4, 1573–1597. MR2440874

[PT17] N. C. Phuc and M. Torres, *Characterizations of signed measures in the dual of* BV *and related isometric isomorphisms*, Ann. Sc. Norm. Super. Pisa Cl. Sci. (5) **17** (2017), no. 1, 385–417. MR3676052

[P] D. Preiss, *Additional regularity for Lipschitz solutions of PDE*, J. Reine Angew. Math. **485** (1997), 197–207. MR1442194

[R] E. Russ, *A survey about the equation* $\operatorname{div} u = f$ *in bounded domains of* $\mathbb{R}^n$, Vietnam J. Math. **41** (2013), no. 4, 369–381. MR3142402

[T] E. Tadmor, *Hierarchical construction of bounded solutions in critical regularity spaces*, Comm. Pure Appl. Math. **69** (2016), no. 6, 1087–1109. MR3493626

[Z] W. P. Ziemer, *Weakly differentiable functions*, Graduate Texts in Mathematics, vol. 120, Springer-Verlag, New York, 1989. Sobolev spaces and functions of bounded variation. MR1014685

Department of Mathematics, Purdue University, West Lafayette, Indiana 47907
Email address: `torres@math.purdue.edu`

Contemporary Mathematics
Volume **709**, 2018
http://dx.doi.org/10.1090/conm/709/14297

A general solution to (free) deterministic equivalents

Carlos Vargas

ABSTRACT. We give an algorithm to compute the asymptotics of the eigenvalue distribution of quite general matricial central limit theorems. The central limits are the so called free deterministic equivalents, which in turn are operators whose Cauchy transforms are the solutions to the equations which define very general deterministic equivalents (a la Girko). Our algorithm is based on the one recently developed by Belinschi, Mai and Speicher and the possibility to extend it to more general, operator-valued situations (in particular, to Benaych-Georges rectangular spaces).

1. Introduction

In this work we survey on the different techniques from free probability which are used to study describe the asymptotic spectrum of a quite large class of random matrices, including those very recently used to model wireless communications (see Chapter 6 of [**17**] or [**1**]). The theory of free probability [**31**] is getting more and more robust and it is quite difficult now to survey on all the aspects that make it such an useful tool for understanding the asymptotic distributions of matrix ensembles.

For this reason, we develop here only those aspects which lead to a quite direct derivation of fixed point equations for computing asymptotic distributions of Hermitian random matrix models.

The general model is described by a (non-commutative) polynomial P evaluated on deterministic matrices and random (Wigner and Haar) matrices. These models were described in [**30**] and were shown to correspond to the solutions of the (recently more recurrent) notion of a deterministic equivalent (DE) for the Cauchy-Stieltjes[1] transform, which go back to Girko [**20**].

Our method is based on properties of Cauchy transforms of operators which can be very neatly described in terms of moments, which in turn are well behaved if we restrict to normal matrices, and in particular, to self-adjoint matrices. We thus require the polynomial to be self-adjoint after being evaluated by a certain tuple of random and deterministic matrices. The tools for the non-selfadjoint case are being

2010 *Mathematics Subject Classification.* Primary 46L54; Secondary 60F05.

Key words and phrases. Free probability, random matrices, deterministic equivalents.

The author was supported by the Deutsche Forschungsgemeinschaft (DFG) through the project SP419/8-1 and by the Austrian Science Foundation (FWF), project P25510-N26.

[1]For a probability measure, we will use the Cauchy transform $G_x(z) := \tau((z-x)^{-1})$, which is just the negative of the Stieltjes-transform $\tau((x-z)^{-1})$.

developed in [**7**]. The matrices that we insert in the polynomial, however, are not required to be self-adjoint. This is one of the main directions in which we extend [**6**], which was up-to-now, the only algorithm general enough to deal with large classes of polynomials. Historically, earlier works were usually devoted to study a specific polynomial (and sometimes even with specific inputs).

A second direction is that we do not ask the input matrices to have a fixed size. We only require that the polynomial multiplies these matrices in such a way that all the summands are of the same size. In particular, the condition on the sizes of the matrices to be proportionally large as $N \to \infty$ (as it is usually assumed when approximating by models by their deterministic equivalent) can be very effectively captured using the formalism of [**9**].

The equations obtained by our method allow to draw the distributions of most of DE's from [**17**]. For some models, such as the Wigner matrices with variance profile, our algorithm fails to be numerically efficient. It is however, theoretically correct and the FDE corresponds to the operator resulting from substituting independent complex gaussians by free circulars, as described already in [**27**]. We should point to another recent, quite general application of operator-valued free probability, which describes the asymptotics of block-modified random matrices in terms of Choi Matrices [**4**] and relies on the free multiplicative convolution [**8**].

In some very broad sense, our method allows to compute Gaussian distributions of a Central limit theorem, for which Wigner's semicircle law is a very special, but fundamental case. We will concentrate here merely on describing such central limits, and not on the qualitative aspects of this convergence.

For more qualitative aspects, such as the analysis of fluctuations, or almost sure convergence (in empirical eigenvalue distributions), a much deeper analysis of moments must be performed. The main aspects of this analysis can already be observed in the proof of Wigner's semicircle law. For this reason, we begin our survey with a sketch of Wigner's semicircle law for the Gaussian unitary ensemble. Then we only point out the main ideas behind the different generalizations, and we refer to the works where such deeper analysis is performed (for the Wigner case in [**25**] and for the Haar case in [**14**]).

Once that the contribution of the basic components (Haar and Wigner matrices and deterministic matrices) is understood, we can apply our knowledge to a specific polynomial to obtain its asymptotic distribution. Although we do not perform this here, a more detailed study of the specific interactions (mixed moments) between the constituting matrices, affects the convergence to the limit. Hence, the polynomial P plays a huge role in this analysis, as it "decides" the rate in which each specific monomial (or cumulant), will appear while computing the moments of P.

For instance, very generally speaking, the self-adjoint polynomials $Z + Z^*$ and ZZ^* have different behaviors when the arguments are replaced by random matrices and their adjoints. The moments of the polynomial ZZ^* are all alternating in Z, which immediately leads to better convergence properties. The shape ZZ^* is recurrent in the models of [**17**] and this explains the better convergence properties that the authors obtain in the original works.

In particular, a deeper analysis of mixed moments allows to understand the fact that in some of these models one is allowed to replace the deterministic matrices by random matrices with bounded operator norms, or by matrices with some (weaker) tightness condition.

1.1. Statement of results. One of most natural ways of constructing a random matrix is to let each entry be an independent copy of a given random variable X. The distribution of X induces a probability measure $\mathbb{P}$ on subsets $E \subseteq M_N(\mathbb{C})$ of matrices. Particularly nice is the matrix $Z = Z_N := (\frac{1}{\sqrt{N}} z_{ij})_{i,j \leq N}$ with independent standard complex Gaussian entries. Such random matrices are called (non-self-adjoint) Gaussian matrices.

From a Gaussian Matrix Z_N there are two immediate ways to a self-adjoint random matrix: The Wigner matrix $X_N = Z_N + Z_N^*$ and the Wishart Matrix $W_N = Z_N Z_N^*$ (where A^* denotes the Hermitian transpose of the matrix A).

Wigner started the study of the asymptotic eigenvalue distribution (AED) of random matrices by establishing the convergence of the AED of X_N to the semicircle law [**36**]. Later, it was shown that such convergence is universal (a kind of Central Limit Theorem), as it holds even if we replace the Gaussian distribution by any other centered distribution (under some mild moment constrains).[2]

Marcenko and Pastur [**24**] studied the second case (and several generalizations of it). They considered first the model $W_N = Z_N Z_N^*$, where Z_N is a $N \times n$ random matrix with independent centered complex Gaussian entries with variance $1/n$. If $N/n \to \lambda \in (0, \infty)$, they showed that the AED of W_N converges to the Marchenko-Pastur law ν which is given by

$$\nu = \begin{cases} (1-\lambda)\delta_0 + \tilde{\nu}, & \text{if } 0 \leq \lambda \leq 1, \\ \tilde{\nu}, & \text{if } 1 < \lambda. \end{cases}$$

where δ_z denotes the Dirac mass at $z \in \mathbb{C}$ and $\tilde{\nu}$ is the measure supported on the interval $[(1-\sqrt{\lambda})^2, (1+\sqrt{\lambda})^2]$, with density

$$\mathrm{d}\tilde{\nu}(t) = \frac{1}{2\pi t}\sqrt{4\lambda - (t-1-\lambda)^2}dt.$$

Equivalently, the measure ν is characterized by its Cauchy-Stieltjes transform G_ν which solves the equation

$$G_\nu(z) = (z - \frac{1}{1-\lambda G_\nu(z)})^{-1}.$$

The Cauchy-Stieltjes transform of a random variable X with distribution μ is defined as

$$G_\mu(z) := G_X(z) := \mathbb{E}((z-X)^{-1}),$$

wherever the inverse of $(z-X)$ exists. If X is supported on the real line, its Cauchy-transform is defined on the whole complex upper half-plane, and its distribution can be recovered from G_X by performing a Stieltjes inversion:

$$\mathrm{d}\mu(t) = \lim_{\varepsilon \downarrow 0} -\frac{1}{\pi}\Im G_X(t+i\varepsilon).$$

As the complexity of the model grows, explicit expressions for the densities quickly become intractable and one can only hope to find equations which determine the Cauchy Transforms of the distributions. In [**24**], they considered also the model $Z_N T_n Z_N^*$, where Z_N is as before and T_n is a self-adjoint deterministic matrix, such that the eigenvalue distribution (μ_{T_n}) of T_n converges to a given probability

[2] To avoid an overwhelming terminology, we will simply call "Wigner Matrices" to both the self-adjoint and the non-self adjoint matrices with centered i.i.d entries and "Gaussian matrices" if the entries are Gaussian. We will distinguish the two cases by using Z for the non-self-adjoint matrices and X for the self-adjoint ones.

measure μ. For this case, the Cauchy transform of the limiting distribution $\nu(\mu)$ satisfies the equation

$$G_{\nu(\mu)}(z) = (z - \int_{\mathbb{R}} \frac{x \mathrm{d}\mu(x)}{1 - x\lambda G_{\nu(\mu)}(z)})^{-1} \tag{1.1}$$

Note that the equations for G_ν and $G_{\nu(\mu)}$ are both non-random and the second equation depends on T_n only through μ. In order to eliminate randomness from the equations, it is crucial that we let $N \to \infty$ (and hence also $n \to \infty$).

For finite N it is much harder to derive the exact eigenvalue distributions. However, the most recent models for wireless communications involve deterministic matrices with fixed finite sizes (which depend, for example, on the number of receiving and transmitting antennas).

The heuristics behind the method of deterministic equivalents (DE) is that, if the matrices involved in the model are large enough, one should still be able to use the asymptotic expressions for the Cauchy transforms to obtain an approximation of the desired distribution.

For example, a DE for the finite dimensional model $Z_N T_n Z_N^*$ is obtained by replacing the limiting deterministic data (λ, μ) by the finite data (Nn^{-1}, μ_{T_n}) in equation (1.1):

$$G_N(z) = (z - \int_{\mathbb{R}} \frac{x \mathrm{d}\mu_{T_n}(x)}{1 - xNn^{-1}G_N(z)})^{-1}.$$

The models $P = \sum_{i=1}^k R_i Z_i T_i Z_i^* R_i^*$ and $Q = \sum_{i=1}^k R_i U_i T_i U_i^* R_i^*$, where the R's and the T's are deterministic and the Z's and U's are, respectively, independent Gaussians and Haar(-distributed unitary) matrices[3], give further generalizations of the Wishart ensemble.

The main contribution of this survey will be an algorithm to approximate the distributions of very general polynomials on deterministic, Gaussian and Haar matrices of different sizes. The models P and Q, are, nevertheless, illustrative enough and they will serve as main examples throughout this work.

In [**15**], it was shown that a DE for the model P is given as the solution of the system of equations:

$$m_N(z) = \frac{1}{N}\mathrm{Tr}(zI_N - \sum_{j=1}^k \int_{\mathbb{R}} \frac{x_j \mathrm{d}\mu_{T_j}(x_j)}{1 - x_j N n_j^{-1} e_j(z)} R_j R_j^*)^{-1} \tag{1.2}$$

where n_j is the size of T_j and

$$e_i = \frac{1}{N}\mathrm{Tr} R_i R_i^* (zI_N - \sum_{j=1}^k \int_{\mathbb{R}} \frac{x_j \mathrm{d}\mu_{T_j}(x_j)}{1 - x_j N n_j^{-1} e_j(z)} R_j R_j^*)^{-1},$$

A similar system of equations was provided for Q in [**16**]. The method of deterministic equivalents was shown to work for some other matrix models (see Chapter 6 of [**17**] for a survey on these). The equations obtained depended in an ad-hoc way on the specific model in question and the models treated are not so diverse (in terms of the non-commutative polynomial on which the model is based).

In [**30**] we proposed a new approach to deterministic equivalents: Instead of considering approximations of the distributions of the matrix models at the level of

[3] We will simply refer to these in the future as "Haar matrices".

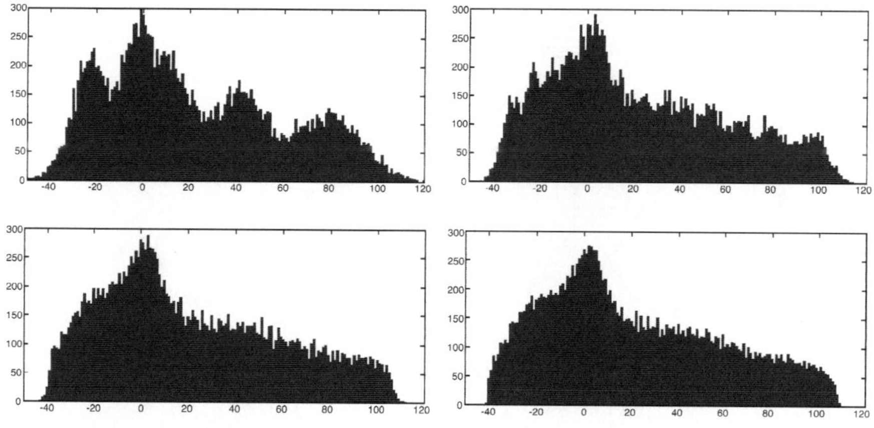

FIGURE 1. 20000 Eigenvalues of Q_m for $Q = Q_1 = R_1U_1T_1U_1^*R_1^* + R_2U_2T_2U_2^*R_2^*$, where R_1, T_1, R_2, T_2 are deterministic matrices of sizes 5×8, 5×4, 8×8 and 4×4 and U_1, U_2 are (chopped) Haar-Unitaries. Here $m = 1$ (up-left), 3 (up-right), 10 (down-left), 40 (down-right).

Cauchy transforms, we approximate the models themselves at the level of operators, inspired by Voiculescu's free probability theory ([**32**], see also [**9**]).

For example, from the matrix models P and Q one can construct the blown-up models

$$P_m = \sum_{i=1}^{k} R_i^{(m)} Z_i^{(m)} T_i^{(m)} (Z_i^{(m)})^* (R_i^{(m)})^*, \quad Q_m = \sum_{i=1}^{k} R_i^{(m)} Z_i^{(m)} T_i^{(m)} (Z_i^{(m)})^* (R_i^{(m)})^*,$$

where the $Z_i^{(m)}$'s (resp. $U_i^{(m)}$'s) are again independent Wigner matrices (resp. Haar matrices) and $A^{(m)} := A \otimes I_m$ for each deterministic matrix A, so that sizes of all the involved matrices are scaled by m.

The collection $(R_1^{(m)}, T_1^{(m)}, Z_1^{(m)}, U_1^{(m)}, \ldots, R_k^{(m)}, T_k^{(m)}, Z_k^{(m)}, U_k^{(m)})$ of blown-up matrices converges (in joint non-commutative distribution as $m \to \infty$) to a very specific collection of operators $(R_1, T_1, c_1, u_1 \ldots, R_k, T_k, c_k, u_k)$ defined in terms of operator-valued free probability. In particular, the AED of P_m and Q_m converge, respectively, to the spectral distribution of the operators $P_\infty := \sum_{i=1}^k R_i c_i T_i c_i^* R_i^*$ and $Q_\infty := \sum_{i=1}^k R_i u_i T_i u_i^* R_i^*$ (see Theorem/Definition 3.1 and Fig. 1.1).

In [**30**] we then used the combinatorial machinery of free probability ([**28**], [**26**]) to show that the DE (e.g. the solution of the equations (similar to 1.2) for Q_N) is exactly the Cauchy-Transform of the *Free Deterministic Equivalent* Q_∞. Moreover, the heuristics behind the method of deterministic equivalents are justified and formalized as a consequence of asymptotic freeness: $Q_m \to Q_\infty$ in distribution, but $Q = Q_1$ is already close to the limit if the involved matrices are large.

The same techniques can be applied for P and for the rest of the models in [**17**] which are obtained by evaluating a fixed non-commutative polynomial on deterministic, Wigner and Haar matrices. The advantage of our approach is that the definition of the FDE can be easily extended to any arbitrary non-commutative

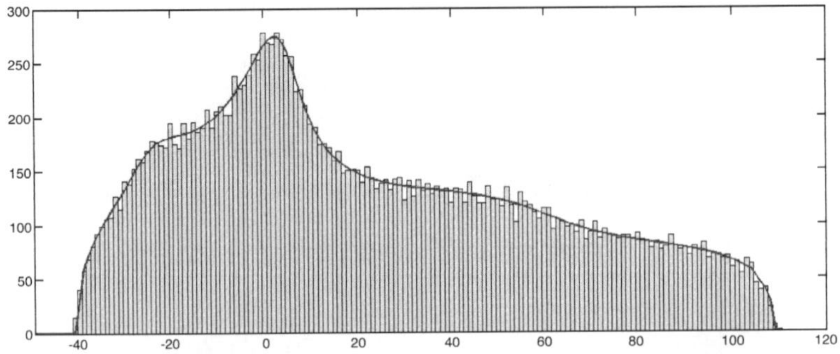

FIGURE 2. 20000 eigenvalues from 100 realizations of Q_{40} (histogram) vs distribution of the free deterministic equivalent (FDE) Q_∞, computed with our algorithm (solid).

polynomial. New models for wireless communications have been proposed using our formalism [**1**]. However, up to know, the actual computation of the distribution of P_∞ was still performed in an ad-hoc way.

In this note, we give a general solution to draw the distribution of a very large class of FDE's (Fig. 2), following the algorithm in [**6**], which in turn relies heavily on the analytic subordination phenomena to deal with operator-valued free convolutions (developed in different instances by Biane [**11**], Voiculescu [**34, 35**], Belinschi and Bercovici [**5**]), and the linearization trick, which was already suggested by Voiculescu, but became quite prominent after the work of Haagerup and Thorbjørnsen ([**21**], see also [**2**]).

Our new algorithm lifts the one in [**6**] to the operator-valued level and relaxes some unnecessary self-adjointness conditions.

This survey is organized as follows: In Section 2 we give a sketch of the proof of Wigner's Theorem. Section 3 includes all the machinery on Operator-valued free probability which is required to formulate our general algorithm and to establish the correspondence between the known DEs and our FDE's. In Section 4 we implement our algorithm to treat several examples (drawn from the literature on wireless communications).

Acknowledgments. The author thanks the free probability work group in Saarbrücken for influential discussions (specially to Tobias Mai and Roland Speicher, for discussions concerning the algorithm in [**6**]). I also thank Franz Lehner and Konrad Schrempf for discussions concerning the linearization trick during my stay in Graz, which helped me to improve the last section of this note.

2. Wigner's theorem

In this Section we sketch the proof of the simplest version of Wigner's semicircle law, namely, the Gaussian case. This will serve us for future reference when dealing with FDE's (which can be thought as very broad generalizations of this fundamental result).

2.1. Moments vs cumulants. Let $X : \Omega \to \mathbb{R}$ be a random variable in a probability space $(\Omega, \mathcal{F}, \mathbb{P})$. The moments of X are the values

$$\mathbb{E}(X^n) = \int_\Omega (X(\omega))^n d\mathbb{P}(\omega).$$

For a large class of random variables, which includes Gaussian random variables and random variables with compact support, the moments of X determine its probability distribution. For several random variables $X_1, \dots, X_k : \Omega \to \mathbb{R}$, we may consider the mixed moments

$$\mathbb{E}(X_1^{n_1} \cdots X_k^{n_k}) = \int_\Omega (X_1(\omega))^{n_1} \cdots (X_k(\omega))^{n_k} d\mathbb{P}(\omega).$$

If $X_1, \dots, X_k$ are determined by their moments, then the stochastic independence of $X_1, \dots, X_k$ is equivalent to the fact that, for all $n_1, \dots, n_k \geq 0$, the mixed moments factorize

$$\mathbb{E}(X_1^{n_1} \cdots X_k^{n_k}) = \mathbb{E}(X_1^{n_1}) \cdots \mathbb{E}(X_k^{n_k}). \tag{2.1}$$

For a collection of random variables $X_1, \dots, X_k$, we define the (multivariate) classical cumulants $K_n(X_{i_1}, \dots X_{i_n})$, $n \leq 1$, $i_1, \dots, i_n \leq k$, recursively as the collection of multi-linear functionals $(K_n)_{\geq 1}$ which satisfy the moment-cumulant formula:

$$\mathbb{E}(X_{i_1} \cdots X_{i_n}) = \sum_{\pi \in \mathcal{P}(n)} K_\pi(X_{i_1}, \cdots, X_{i_n}),$$

where:

DEFINITION 2.1. *$\mathcal{P}(n) := \mathcal{P}([n])$ are the set partitions; each element $\pi = \{V_1, \dots, V_{|\pi|}\} \in \mathcal{P}([n])$ decomposes the set $[n] = V_1 \cup V_2 \cup \dots \cup V_{|\pi|}$ into non-empty, pairwise disjoint subsets ("blocks") $V_1, \dots, V_{|\pi|}$. We write*

$$K_\pi(X_1, \dots, X_n) := \prod_{V \in \pi} K_{|V|}(X_V),$$

where we use the notation

$$K_{|V|}(X_V) := K_m(X_{v_1}, \dots, X_{v_m})$$

for each block $V = \{v_1, \dots, v_m\} \in \pi$, $v_1 < \dots < v_m$.

For example, $K_1(X_i) = \mathbb{E}(X_i)$ is simply the mean and $K_2(X_i, X_j) = \mathbb{E}(X_i X_j) - \mathbb{E}(X_i)\mathbb{E}(X_j)$ is the covariance.

For an ordered tuple $\mathbf{X} = (X_1, \dots, X_k)$ of random variables, we call

$$\Phi_m^{\mathbf{X}} := \Phi_m = \{(i_1, \dots, i_m) \mapsto \mathbb{E}(X_{i_1} \dots X_{i_m}) : i_1, \dots, i_m \leq k\}$$

the m-th order mixed moments of $(X_1, \dots, X_k)$. Analogously we define

$$\Psi_m^{\mathbf{X}} := \{(i_1, \dots, i_m) \mapsto K_m(X_{i_1}, \dots, X_{i_m}) : i_1, \dots, i_m \leq k\}$$

the m-th order cumulants. The collection of moment maps $(\Phi_m^{\mathbf{X}})_{m \leq n}$ contains exactly the same information as the collection of mixed cumulants $(\Psi_m^{\mathbf{X}})_{m \leq n}$. However, cumulants seem to organize statistical information in a nicer way.

A real random variable X is the constant random variable $X = c \in \mathbb{R}$ iff all cumulants of degree $n \geq 2$ vanish and $\kappa_1(X) = c$. In fact, it is not hard to see that, if we input a constant in any of the arguments of a cumulant of order $k \geq 2$, then the cumulant must vanish, independently from the position of the constant argument and the rest of the arguments.

In terms of cumulants, the simplest (non-constant) random variables are the Gaussian random variables: X has the Normal distribution $\mathcal{N}(\mu, \sigma^2)$ iff all cumulants of degree $n \geq 3$ vanish, $K_2(X, X) = \sigma^2$ and $K_1(X) = \mu$.

For two independent random variables X, Y, the covariance must vanish. In fact, the higher order cumulants are generalizations of the covariance, and the relation of X, Y being independent holds if and only if all the mixed cumulants (i.e. $K_2(X,Y), K_2(Y,X), K_3(X,X,Y), K_3(X,Y,X), K_3(X,Y,Y), \dots$ etc.) vanish. This implies in particular that the cumulants of $X+Y$ are simply

$$K_n(X+Y, X+Y, \dots, X+Y) = K_n(X, \dots, X) + K_n(Y, \dots, Y).$$

Hence, cumulants can be used to compute additive convolutions (in fact, the cumulants are related to the coefficients of Fourier transforms).

If X, Y are independent standard Gaussian ($\mathcal{N}(0,1)$) random variables, then $Z := (X + iY)/\sqrt{2}$ has the standard complex Gaussian distribution, which can also be characterized in terms of the mixed cumulants of Z and $Z^* := \bar{Z}$. The only non-vanishing cumulants are

$$K_2(Z, Z^*) = K_2(Z^*, Z) = 1.$$

Wick's formula for independent complex Gaussians $Z_1, \dots, Z_k$, states that, for any $\varepsilon = (\varepsilon_1, \dots, \varepsilon_n) \in \{1, *\}^n$, we have

$$\mathbb{E}(Z_{i_1}^{\varepsilon_1} \dots Z_{i_n}^{\varepsilon_n}) = \sum_{\pi \in \mathcal{P}_2(n)} K_\pi(Z_{i_1}^{\varepsilon_1}, \dots, Z_{i_n}^{\varepsilon_n}), \tag{2.2}$$

where $\mathcal{P}_2(n) \subset \mathcal{P}(n)$ denotes the subset of matchings (i.e. partitions $\pi \in \mathcal{P}(n)$ such that every $V \in \pi$ has exactly 2 elements). We observe that the Wick formula for this case can be easily derived from the moment-cumulant formula: the restriction to pairings $\mathcal{P}_2(n) \subset \mathcal{P}(n)$ follows from the fact that only the second order cumulants of standard complex Gaussians may not vanish. Furthermore, any block (r, s) of π should not match independent variables (since the mixed cumulant, and hence K_π would vanish).

However, going over cumulants to compute $\mathbb{E}(Z_{i_1}^{\varepsilon_1} \dots Z_{i_n}^{\varepsilon_n})$ seems more like a detour in this case. We could simply compute $\mathbb{E}(Z_{i_1}^{\varepsilon_1} \dots Z_{i_n}^{\varepsilon_n})$ by a direct application of the factorization in eq. (2.1). Our Wick formula (2.2) will be very useful when we go over to random matrices.

2.2. Gaussian matrices and Wigner's semicircle law. Let $Z = Z_N := (\frac{1}{\sqrt{N}} z_{ij})_{i,j \leq N}$ be a matrix with independent standard complex Gaussian entries $(z_{ij})_{i,j \leq N}$ (the choice of the normalization $\frac{1}{\sqrt{N}}$ will be clear later).

In **[36]** Wigner described the asymptotic eigenvalue distribution (as $N \to \infty$) of the (necessarily real) eigenvalues of $X_N = (Z_N + Z_N^*)/\sqrt{2}$ (where A^* denotes the Hermitian transpose of A).

Since $X(\omega) = (X(\omega))^*$ for any realization of X, we can diagonalize

$$X(\omega) = U(\omega) D(\omega) (U(\omega))^*,$$

where $D = diag(\lambda_1(\omega), \dots, \lambda_N(\omega))$ and hence for all $k \geq 0$, we have

$$\begin{aligned}\frac{1}{N}\mathrm{Tr}(X^k(\omega)) = \frac{1}{N}\mathrm{Tr}((U(\omega)D(\omega)(U(\omega))^*)^k) &= \frac{1}{N}\mathrm{Tr}(U(\omega)D(\omega)^k(U(\omega))^*)\\ &= \frac{1}{N}\mathrm{Tr}(D(\omega)^k)\\ &= \frac{1}{N}\sum_{i\leq N}(\lambda_i(\omega))^k.\end{aligned}$$

For each ω, expression $\frac{1}{N}\sum_{i\leq N}(\lambda_i(\omega))^k$ can be identified as the k-th moment of the real random variable $\Lambda(X(\omega))$, with discrete probability measure $\mu_{X(\omega)}$ which assigns a mass of $1/N$ to each eigenvalue of $X(\omega)$. The averaged eigenvalue distribution (AED) μ_X is the distribution of the random variable $\Lambda(X)$ obtained by averaging all such $\Lambda(X(\omega))$, $\omega \in \Omega$ against $\mathbb{P}$. More specifically, it is the probability measure μ_{X_N} with k-th moment equal to $\frac{1}{N}\mathbb{E}(\sum_{i\leq N}\lambda_i^k)$. We want to describe μ_{X_N} when the size $N \to \infty$. Let us denote $\tau_N := \frac{1}{N}\mathrm{Tr}$.

Instead of computing the (rather complicated) joint distributions of $(\lambda_1, \lambda_2, \dots, \lambda_N)$, we compute the moments

$$\mathbb{E}\circ\tau_N(X^k(\omega)) = \frac{1}{N}\mathbb{E}(\sum_{i\leq N}\lambda_i^k),$$

which depend on the entries of our matrices in a polynomial way, hoping that we will be able to identify them as moments of a probability measure. Indeed, this will be the case, and for this, it will be convenient to work first directly with Z and Z^* and symmetrizing only at the very end.

In general, we would like to compute, for all $k \geq 1$ and every $\varepsilon = (\varepsilon_1, \dots, \varepsilon_n) \in \{1, *\}^k$

$$\mathbb{E}\circ\tau_N(Z^{\varepsilon_1}, \dots, Z^{\varepsilon_k}).$$

As an example, let us consider the fourth order mixed moment $\varepsilon = (1, *, *, 1)$, we have

$$\mathbb{E}\circ\tau_N(ZZ^*Z^*Z) = \frac{1}{N^3}\sum_{i_1,\dots,i_4\leq N}\mathbb{E}(z_{i_1i_2}\bar{z}_{i_3i_2}\bar{z}_{i_4i_3}z_{i_4i_1}) \tag{2.3}$$

By Wick's Formula,

$$\mathbb{E}(z_{i_1i_2}\bar{z}_{i_3i_2}\bar{z}_{i_4i_3}z_{i_3i_1}) = \sum_{\pi\in\mathcal{P}_2(4)}K_\pi(z_{i_1i_2}, \bar{z}_{i_3i_2}, \bar{z}_{i_4i_3}, z_{i_4i_1}).$$

There are 3 pairings of $\{1,2,3,4\}$, namely $\pi_1 = \{\{1,2\}\{3,4\}\}$, $\pi_2 = \{\{1,3\}\{2,4\}\}$, $\pi_3 = \{\{1,4\}\{2,3\}\}$. Since the z_{ij}'s are complex Gaussian random variables, π_3 will vanish (independently of the choice of $i_1, \dots, i_4$) since it will never match a z_{ij} with $\bar{z}_{ij}$, which is a necessary condition for the cumulant not to vanish. Hence

$$\mathbb{E}\circ\tau_N(ZZ^*Z^*Z) = \frac{1}{N^3}\sum_{i_1,\dots,i_4\leq N}K_{\pi_1}(z_{i_1i_2}, \bar{z}_{i_3i_2}, \bar{z}_{i_4i_3}, z_{i_4i_1}) \tag{2.4}$$

$$+ \frac{1}{N^3}\sum_{i_1,\dots,i_4\leq N}K_{\pi_2}(z_{i_1i_2}, \bar{z}_{i_3i_2}, \bar{z}_{i_4i_3}, z_{i_4i_1}) \tag{2.5}$$

Since the entries are independent standard complex Gaussians, each partition imposes restrictions on the indices for the cumulants not to vanish, namely

$$\begin{aligned} K_{\pi_1}(z_{i_1i_2}, \bar{z}_{i_3i_2}, \bar{z}_{i_4i_3}, z_{i_4i_1}) = K_2(z_{i_1i_2}, \bar{z}_{i_3i_2})K_2(\bar{z}_{i_4i_3}, z_{i_4i_1}) &= \delta_{i_1i_3}\delta_{i_3i_1}, \\ K_{\pi_2}(z_{i_1i_2}, \bar{z}_{i_3i_2}, \bar{z}_{i_4i_3}, z_{i_4i_1}) = K_2(z_{i_1i_2}, \bar{z}_{i_4i_3})K_2(\bar{z}_{i_3i_2}, z_{i_4i_1}) &= \delta_{i_1i_4}\delta_{i_3i_2}\delta_{i_3i_4}\delta_{i_2i_1}, \end{aligned}$$

Hence we need only to count the number of free indices in order to obtain the contribution of each partition. For this case we obtain

$$\mathbb{E} \circ \tau_N(ZZ^*Z^*Z) = 1 + 1/N^2.$$

As $N \to \infty$ only the contribution of π_1 will survive. For a general moment of order k, an easy inductive argument shows that a pairing $\pi \in \mathcal{P}_2(k)$ can only contribute in the limit if $\pi \in \mathcal{NC}_2(k) \subset \mathcal{P}_2(k)$ is a non-crossing pairing (i.e. there is no quadruple $1 \leq a < b < c < d \leq k$ such that $a, c \in V_i, b, d \in V_j$ where $V_i \neq V_j$ are blocks of π). In addition, we must have that, for each matching $\{a, b\} \in \pi$, $\varepsilon_a \neq \varepsilon_b$ (in contrast to π_3 above). Hence, for computing the asymptotics of a general moment we need to find

$$\frac{1}{N^{1+k/2}} \sum_{\substack{i_1,\ldots,i_k \leq N \\ \pi \in \mathcal{NC}_2(k)}} K_\pi(z_{i_1i_2}^{\varepsilon_1}, z_{i_2i_3}^{\varepsilon_2}, \ldots, z_{i_ki_1}^{\varepsilon_k}) = |\mathcal{NC}_\varepsilon(k)|, \tag{2.6}$$

where $\pi \in \mathcal{NC}_\varepsilon(k) \subseteq \mathcal{NC}_2(k)$ iff $\varepsilon_a \neq \varepsilon_b$ for each pair $\{a, b\} \in \pi$.

Now, since $X = (Z + Z^*)/\sqrt{2}$, we have that

$$\lim_{N\to\infty} \mathbb{E} \circ \tau_N(X^k) = \frac{1}{N^{1+k/2}2^{k/2}} \sum_{\substack{i_1,\ldots,i_k \leq N \\ \varepsilon=(\varepsilon_1,\ldots,\varepsilon_k)\in\{1,*\}^k \\ \pi \in \mathcal{NC}_2(k)}} K_\pi(z_{i_1i_2}^{\varepsilon_1}, z_{i_2i_3}^{\varepsilon_2}, \ldots, z_{i_ki_1}^{\varepsilon_k}).$$

If we fix a non-crossing pairing $\pi \in \mathcal{NC}_2(k)$, there are $2^{k/2}$ non-vanishing choices for ε (for each block of $(a, b) \in \pi$, we can have $(\varepsilon_a, \varepsilon_b)$ equal to either $(1, *)$ or $(*, 1)$). Hence, after summing over all free indices and all $\varepsilon \in \{1, *\}^k$, each non-crossing pairing contributes with 1 to the moment in the limit. It is well-known that the number of non-crossing pairings $|\mathcal{NC}_2(k)|$ are counted by the Catalan numbers $C_n := \frac{1}{n+1}\binom{n}{2n}$, which in turn, are the even moments of the semicircular distribution supported on $[-2, 2]$ with density

$$\mathrm{d}\mu(t) = \frac{1}{2\pi}\sqrt{4 - t^2}.$$

Hence the assertion follows.

In the next section we introduce Voiculescu's free probability theory for non-commutative random variables, which allows to treat random matrices as random variables. One of the key ideas is to think of $\tau_N = \frac{1}{N}\mathbb{E} \circ \mathrm{Tr}$ as a generalization of the expectation $\mathbb{E}$ in classical probability.

3. Operator-valued free probability

Voiculescu's free probability [**31**] is a prominent branch of non-commutative probability, where the classical probabilistic notion of moments (with respect to the expectation) is extended to more general linear functionals $\tau : \mathcal{A} \to \mathbb{C}$ on a C^*-(or W^*-)algebra $\mathcal{A}$. A crucial step in the development of free probability was to further extend this to cover the classical notion of conditional expectation [**33**].

DEFINITION 3.1. *Let $\mathcal{A}$ be a unital $*$-algebra and let $\mathbb{C} \subseteq \mathcal{B} \subseteq \mathcal{A}$ be a $*$-sub-algebra. A $\mathcal{B}$-probability space is a pair $(\mathcal{A}, \mathbf{F})$, where $\mathbf{F} : \mathcal{A} \to \mathcal{B}$ is a conditional expectation, that is, a linear map satisfying:*

$$\begin{aligned} \mathbf{F}(bab') &= b\mathbf{F}(a)b', \qquad \forall b, b' \in \mathcal{B}, a \in \mathcal{A} \\ \mathbf{F}(1) &= 1. \end{aligned}$$

The elements $a \in \mathcal{A}$ are called (non-commutative) random variables and the pair $(\mathcal{A}, \mathbf{F})$ is called a $\mathcal{B}$-valued probability space. The situation $\mathcal{B} = \mathbb{C}$ is usually highlighted by writing τ instead of $\mathbf{F}$ and the pair $(\mathcal{A}, \tau)$ is then called a (scalar-valued) non commutative probability space (NCPS).

The main virtue of Definition 3.1 is that we can treat simultaneously classical random variables and deterministic matrices in the same framework. These spaces are important building blocks for many interesting examples.

EXAMPLES 3.2. *(1). Complex-valued random variables in a classical probability space $(\Omega, \mathcal{F}, \mathbb{P})$ can be thought as a NCPS $(\mathcal{A}, \tau)$ with involution given by complex conjugating the random variables and $\tau := \mathbb{E}$ being the usual expectation. Any sub-sigma algebra $\mathcal{H} \subset \mathcal{F}$ induces a conditional expectation. This means a map $X \mapsto \mathbb{E}(X|\mathcal{H})$, where for any $E \in \mathcal{H}$,*

$$\int_E \mathbb{E}(X|\mathcal{H}) \mathrm{d}\mathbb{P} = \int_E X \mathrm{d}\mathbb{P}.$$

In particular for the trivial sub-sigma algebra $\mathcal{H}_1 = \{\emptyset, \Omega\}$ we recover the usual expectation and for $\mathcal{H}_2 = \mathcal{F}$ we obtain the identity map $\mathbb{E}(X|\mathcal{H}_2) = X$.

(2) The NCPS $(\mathcal{A} := \mathcal{M}_N(\mathbb{C}), \tau_N := \frac{1}{N}\mathrm{Tr})$ of complex $N \times N$ matrices with the normalized trace is a $*$-probability space (with involution given by the Hermitian transpose). As we saw in the previous section, if $X \in \mathcal{M}_N(\mathbb{C})$ is Hermitian, we may diagonalize $X = UDU^*$ and realize that

$$\tau(X^k) = \tau(UD^kU^*) = \tau(D^k) = \frac{1}{N}\sum_{i=1}^{N} \lambda_i(X)^k = \int_{\mathbb{R}} x^k \mu_X(dt),$$

where $\lambda_1(X) \leq \cdots \leq \lambda_N(X)$ are the eigenvalues of X and $\mu_X = \frac{1}{N}\sum_{i=1} \delta_{\lambda_i(X)}$.

Hence $(\tau(a^k))_{k \geq 1}$ are the moments (in the usual, probabilistic sense) of the averaged eigenvalue distribution μ_X. Such a correspondence between moments and probability distributions (on the complex plane) works not only for Hermitian matrices but for general normal matrices (which in particular include all Hermitian, Skew-Hermitian and Unitary matrices).

Combining (via the algebraic tensor product) a classical probability space $(\mathcal{A}, \mathbb{E})$ with the NCPS of deterministic matrices $(\mathcal{M}_N(\mathbb{C}), \tau_N)$ yields the NCPS $(\mathcal{A} \otimes \mathcal{M}_N(\mathbb{C}), \mathbb{E} \otimes \tau_N)$, which can be identified with the space of matrices with random entries drawn from $\mathcal{A}$. In this way, random matrices can be treated as non commutative random variables.

EXAMPLES 3.3. *Let $(\mathcal{A}, \tau)$ be a NCPS.*

(1). Let $p_1, \ldots, p_k \in \mathcal{A}$ be pairwise orthogonal projections with $1_{\mathcal{A}} = p_1 + \cdots + p_k$. There exist a unique conditional expectation $\mathbf{F} : \mathcal{A} \to \langle p_1, \ldots, p_k \rangle$ compatible with τ in the sense that $\tau \circ \mathbf{F} = \tau$, explicitly given by

$$\mathbf{F}(a) = \sum_{i \leq k} \tau(p_i)^{-1} \tau(p_i a p_i).$$

(2). Consider the algebra $M_N(\mathcal{A}) \cong M_N(\mathbb{C}) \otimes \mathcal{A}$ of $N \times N$ matrices with entries in $\mathcal{A}$. The maps

$$\mathbf{F}_3 : (a_{ij})_{ij} \mapsto (\tau(a_{ij}))_{ij} \in M_N(\mathbb{C}), \tag{3.1}$$

$$\mathbf{F}_2 : (a_{ij})_{ij} \mapsto (\delta_{ij}\tau(a_{ij}))_{ij} \in D_N(\mathbb{C}), \tag{3.2}$$

$$\mathbf{F}_1 : (a_{ij})_{ij} \mapsto \sum_{i=1}^{N} \frac{1}{N}\tau(a_{ii})I_N \in \mathbb{C} \cdot I_N \tag{3.3}$$

are respectively, conditional expectations onto the algebras $M_n(\mathbb{C}) \supset D_n(\mathbb{C}) \supset \mathbb{C}\cdot I_N$ of constant matrices, diagonal matrices and multiples of the identity. Note that $(M_N(\mathbb{C}) \otimes \mathcal{A}, \mathbf{F}_1)$ is a scalar-valued NCPS.

The generality of non-commutative random variables allowed Voiculescu to define a parallel notion to (conditional) independence, which he called *freeness* (*with amalgamation*).

DEFINITION 3.4. *Let $(\mathcal{A}, \mathbf{F})$ be a $\mathcal{B}$-probability space and let $\bar{a} := a - \mathbf{F}(a)1_{\mathcal{A}}$ for any $a \in \mathcal{A}$. The $*$-sub-algebras $\mathcal{B} \subseteq A_1, \dots, A_k \subseteq \mathcal{A}$ are $\mathcal{B}$-free (or free over $\mathcal{B}$, or free with amalgamation over $\mathcal{B}$) (with respect to $\mathbf{F}$) iff*

$$\mathbf{F}(\bar{a_1}\bar{a_2}\cdots\bar{a_m}) = 0, \tag{3.4}$$

for all $m \geq 1$ and all tuples $a_1, \dots, a_m \in \mathcal{A}$ such that $a_i \in A_{j(i)}$ with $j(1) \neq j(2) \neq \cdots \neq j(m)$ (note that, for example, $j(1) = j(3)$ is allowed).

Subsets $S_1, \dots, S_k \subset \mathcal{A}$ are $\mathcal{B}$-free if so are the $*$-sub-algebras $\langle S_1, \mathcal{B}\rangle, \dots, \langle S_k, \mathcal{B}\rangle$.

Free probability is a realm, parallel to classical probability, where the factorization of expectations (Eq. 2.1) given by classical independence is replaced by freeness. Many fundamental theorems from classical probability, such as the convergence to the Central Limit or the Law of small numbers can be translated to the free setting.

In particular, by just replacing independence by free independence and working algebraically, the free Central Limit (i.e. the limiting distribution of $S_N = \frac{1}{\sqrt{N}}(\bar{a}_1 + \cdots + \bar{a}_N)$ for free self-adjoint, identically distributed elements $(a_i)_{i\geq 1})$ turns out to be Wigner's semicircle law, whereas the free analog of the law of small numbers is the Marchenko-Pastur distribution (also known as free Poisson), which is the (also universal) limit of singular-value distributions of Wishart matrices.

In his seminal work [**32**], Voiculescu constructed free operators inspired by these pioneering results on the asymptotics of random matrices. In order to state his results on asymptotic freeness, we need to define special non-commutative random variables, and the notion of convergence in non-commutative distribution.

DEFINITION 3.5. *Let $(\mathcal{A}, \tau)$ be a NCPS.*

(1). A random variable $c \in \mathcal{A}$ is a circular element iff, for any $k \geq 1$ and $\varepsilon = (\varepsilon_1, \dots, \varepsilon_k) \in \{1, *\}^k$, the mixed moment on (c, c^*) are

$$\tau(c^{\varepsilon_1}c^{\varepsilon_2}\dots c^{\varepsilon_k}) = |\mathcal{NC}_\varepsilon(k)|.$$

(2). A random variable $u \in \mathcal{A}$ is a Haar-unitary iff $u^*u = 1 = uu^*$ and $\tau(u^k) = 0$ for all $k \geq 1$.

Note that c is not a normal operator, (for example, $\tau(ccc^*c^*) = 1 \neq 2 = \tau(cc^*cc^*)$ which would be equal if $c^*c = cc^*$). However, by Equation 2.7 (and the discussion afterwards), it is easy to see that the self-adjoint random variable $s = 2^{-1/2}(c + c^*)$ has the standard semicircular distribution.

DEFINITION 3.6. *(1). For an ordered tuple $a = (a_1, \ldots, a_k)$ of random variables, we define the distribution of a as the collection $\Phi(a) := (\Phi_m)_{m\geq 1}$ of maps*

$$\Phi_m : \{(i_1, \ldots, i_m) \mapsto \tau(a_{i_1} \ldots a_{i_m}) : i_1, \ldots, i_m \leq k\},$$

If $a = (a_1, \ldots, a_k) \in \mathcal{A}_1^k$ and $b = (b_1, \ldots, b_k) \in \mathcal{A}_2^k$ are tuples of random variables in (possibly different) NCPS $(\mathcal{A}_1, \tau_1)$, $(\mathcal{A}_2, \tau_2)$, such that $\tau_1(a_{i_1} \ldots a_{i_m}) = \tau_2(b_{i_1} \ldots b_{i_m})$ for all $m \geq 1$ and all $1 \leq i_1, \ldots, i_m \leq k$, we say that a and b have the same distribution and we write $a \sim b$

(2). Let $(\mathcal{A}_N, \tau_N)$, $N \geq 1$, and $(\mathcal{A}, \tau)$ be $\mathbb{C}$-probability spaces and let $(a_1^{(N)}, \ldots, a_k^{(N)}) \in \mathcal{A}_N^k$, $(a_1, \ldots, a_k) \in \mathcal{A}^k$ be such that

$$\lim_{N\to\infty} \tau_N((a_{i_1}^{(N)}) \cdots (a_{i_r}^{(N)})) = \tau(a_{i_1} \cdots a_{i_r}),$$

for all $r \geq 1$, $1 \leq i_1, \ldots, i_m \leq k$ Then we say that $(a_1^{(N)}, \ldots, a_k^{(N)})$ converges in distribution to $(a_1, \ldots, a_k)$ and we write $(a_1^{(N)}, \ldots, a_k^{(N)}) \to (a_1, \ldots, a_k)$.

3.1. Asymptotic freeness of random matrices.

THEOREM 3.7. *For each $N \geq 1$, let $Z_1^{(N)}, \ldots, Z_p^{(N)}$ and $U_1^{(N)}, \ldots, U_q^{(N)}$ be $N \times N$ independent Wigner and Haar unitary matrices. Let $D_1^{(N)}, \ldots, D_r^{(N)}$ be deterministic matrices, such that, for any $k \geq 1$ and $1 \leq j_1, \ldots, j_k \leq r$ there exist a constant $c(j_1, \ldots, j_k) \in \mathbb{C}$ such that*

$$\lim_{N\to\infty} \frac{1}{N} \operatorname{Tr}((D_{j_1}^{(N)})(D_{j_2}^{(N)}) \cdots (D_{j_k}^{(N)})) = c(j_1, \ldots, j_k). \tag{3.5}$$

Then, as $N \to \infty$,

$$(X_1^{(N)}, \ldots, X_p^{(N)}, U_1^{(N)}, \ldots, U_q^{(N)}, D_1^{(N)}, \ldots, D_r^{(N)}) \to (s_1, \ldots, s_p, u_1, \ldots u_q, d_1, \ldots d_r)$$

where $s_1, \ldots, s_p, u_1, \ldots u_q, d_1, \ldots d_r$ are elements in some $*$-probability space $(\mathcal{A}, \tau)$ whose joint-distribution is determined by the following conditions:

- *s_i is a semicircular element for all $i \leq p$.*
- *u_i is a Haar-unitary for all $i \leq q$.*
- *$\tau(d_{i_1} d_{i_2} \cdots d_{i_k}) = c(i_1, \ldots, i_k)$, for any $k \geq 1$, $1 \leq i_1, \ldots, i_k \leq r$.*
- *The algebras $\langle s_1 \rangle, \ldots, \langle s_p \rangle, \langle u_1, u_1^* \rangle, \ldots, \langle u_q, u_q^* \rangle, \langle d_1, \ldots, d_r \rangle$ are ($\mathbb{C}$-)free.*

Theorem 3.7 generalizes Wigner's semicircle law in several directions. It allows us to compute the asymptotic mixed moments of

$$(X_1^{(N)}, \ldots, X_p^{(N)}, U_1^{(N)}, \ldots, U_q^{(N)}, D_1^{(N)}, \ldots, D_r^{(N)})$$

by means of the rule of free independence (3.4), in terms of the individual asymptotic moments of $X_1^{(N)}, \ldots, X_p^{(N)}, U_1^{(N)}, \ldots, U_q^{(N)}$ and the (given) asymptotic mixed moments of $(D_1^{(N)}, \ldots, D_r^{(N)})$. A nice way to understand how these mixed moments are calculated is in terms of free cumulants (see Section 3.3).

Our combinatorial proof of the Gaussian case in Section 2 will be our main reference to indicate how the different generalizations work.

- Relaxing Gaussian condition.

Let us consider first a single matrix Z. Again, we need to study

$$\frac{1}{N}\mathbb{E}\circ\mathrm{Tr}(Z^{\varepsilon_1},\dots,Z^{\varepsilon_k})=\frac{1}{N^{1+k/2}}\sum_{\substack{i_1,\dots,i_k\leq N\\ \pi\in\mathcal{P}(k)}}K_\pi(z^{\varepsilon_1}_{i_1i_2},z^{\varepsilon_2}_{i_2i_3},\dots,z^{\varepsilon_k}_{i_ki_1}). \tag{3.6}$$

For the Gaussian case all cumulants of order different than 2 vanished and hence our sum ran over $\mathcal{P}_2(k)$. Then we noticed that only $\pi\in\mathcal{NC}_2(k)$ actually matter in the limit.

If Z is no longer Gaussian, we need to consider partitions $\pi\in\mathcal{P}(k)$ such that all blocks of π are of size greater or equal to 2, but $\pi\in\mathcal{P}(k)$ needs not in principle to be in $\mathcal{P}_2(k)$. Blocks of size 1 are still not allowed because the entries of Z are all centered. The important observation is that the number of different cumulants to be considered depends only on the size k of the moment in question (and does not grow with N, which, for each π, only affects the number of choices for each free index).

If we allow π to have bigger blocks, it is intuitive that the contribution of π will vanish in the limit since it will imply that more indices need to be identified. This can be shown by induction.

- Several Wigner Matrices.

Going from one Wigner matrix to several is not hard. We now need to consider the more general expression, for $j_1,\dots,j_k\leq p$

$$\frac{1}{N}\mathbb{E}\circ\mathrm{Tr}(Z^{\varepsilon_1}_{j_1},\dots,Z^{\varepsilon_k}_{j_k})=\frac{1}{N^{1+k/2}}\sum_{\substack{i_1,\dots,i_k\leq N\\ \pi\in\mathcal{P}(k)}}K_\pi(z^{\varepsilon_1;j_1}_{i_1i_2},z^{\varepsilon_2;j_2}_{i_2i_3},\dots,z^{\varepsilon_k;j_k}_{i_ki_1}),$$

where now the blocks of $\pi\in\mathcal{P}(n)$ should also respect the labels indicated by $j=(j_1,\dots,j_k)$. Such restriction can be then carried out to the very end. In the limit, the sum will again run over non-crossing pairings $\pi\in\mathcal{NC}_\varepsilon(n)\subset\mathcal{NC}_2(n)$ with the additional condition that for all $(a,b)\in\pi$, $j_a=j_b$. We will see later in Section 3.3 that this characterizes a free family of circular operators.

- Wigner matrices and deterministic matrices.

If we now allow deterministic matrices to operate between our Wigner matrices we need to compute expressions of the form

$$\frac{1}{N}\mathbb{E}\circ\mathrm{Tr}(D_{j_0}Z^{\varepsilon_1}_{j_1}D_{j_2},\dots,Z^{\varepsilon_k}_{j_{2k-1}}D_{j_{2k}}).$$

More generally, we are interested in estimating

$$\frac{1}{N^{1+k/2}}\sum_{\substack{i_0,\dots,i_{2k+1}\leq N\\ \pi\in\mathcal{P}(2k+1)}}K_\pi(d^{(j_0)}_{i_0i_1},z^{\varepsilon_1}_{i_1i_2},d^{(j_2)}_{i_2i_3},z^{\varepsilon_2}_{i_3i_4},\dots,z^{\varepsilon_k}_{i_{2k-1}i_{2k}},d^{(j_{2k})}_{i_{2k}i_{2k+1}}).$$

Since the d_{ij}'s are constants, $\{\{1\},\{3\},\dots,\{2k+1\}\}\subset\pi$, otherwise the cumulant vanishes. Therefore, we need only to consider partitions of $\mathcal{P}(2,4,\dots,2k)\cong\mathcal{P}(k)$.

As an example, let us assume that we only have a single Gaussian matrix and consider the pairing $\{\{1,3\}\{2,7\}\{4,5\}\{6,8\}\}$ and $\varepsilon=(1,*,*,1,*,1,1,*)$. Again, each block corresponds to a cumulant which imposes identifications of some indices:

$$h_0:=i_0,\quad h_1:=i_1=i_6,\quad h_2:=i_2=i_5,\quad h_3:=i_4=i_{13},\quad h_4:=i_3=i_{14},$$
$$h_5:=i_7=i_{10},\quad h_6:=i_8=i_9,\quad h_7:=i_{11}=i_{16},\quad h_8:=i_{12}=i_{15},\quad h_9:=i_{17}$$

The contribution of π will be $K_\pi(z,\bar{z},\bar{z},z,\bar{z},z,z,\bar{z})$ (which is one in this case, and does not depend on N in general), times $\mathrm{Tr}(D_{\pi,\varepsilon})$, where

$$D_{\pi,\varepsilon}=\sum_{h_0,\dots,h_9\le N} d^{(j_0)}_{i_0i_1}d^{(j_2)}_{i_2i_3}\cdots d^{(j_{2k})}_{i_{2k}i_{2k+1}}.$$

For our example the sum $D_{\pi,\varepsilon}$ is given by:

$$\begin{aligned}
(3.7)\quad D_{\pi,\varepsilon} &= \sum_{h_0,\dots,h_9\le N} d^{(j_0)}_{i_0i_1}d^{(j_2)}_{i_2i_3}d^{(j_4)}_{i_4i_5}d^{(j_6)}_{i_6i_7}d^{(j_8)}_{i_8i_9}d^{(j_{10})}_{i_{10}i_{11}}d^{(j_{12})}_{i_{12}i_{13}}d^{(j_{14})}_{i_{14}i_{15}}d^{(j_{16})}_{i_{16}i_{17}}\\
(3.8)\quad &= \sum_{h_0,\dots,h_9\le N} d^{(j_0)}_{h_0h_1}d^{(j_2)}_{h_2h_4}d^{(j_4)}_{h_3h_2}d^{(j_6)}_{h_1h_5}d^{(j_8)}_{h_6h_6}d^{(j_{10})}_{h_5h_7}d^{(j_{12})}_{h_8h_3}d^{(j_{14})}_{h_4h_8}d^{(j_{16})}_{h_7h_9}\\
(3.9)\quad &= \sum_{h_0,\dots,h_9\le N} (d^{(j_0)}_{h_0h_1}d^{(j_6)}_{h_1h_5}d^{(j_{10})}_{h_5h_7}d^{(j_{16})}_{h_7h_9})(d^{(j_4)}_{h_3h_2}d^{(j_2)}_{h_2h_4}d^{(j_{14})}_{h_4h_8}d^{(j_{12})}_{h_8h_3})(d^{(j_8)}_{h_6h_6})\\
(3.10)\quad &= D_{j_0}D_{j_6}D_{j_{10}}D_{j_{16}}\mathrm{Tr}(D_{j_4}D_{j_2}D_{j_{14}}D_{j_{12}})\mathrm{Tr}(D_{j_8}),
\end{aligned}$$

Hence

$$\begin{aligned}
\frac{1}{N^{1+k/2}}\mathrm{Tr}(D_{\pi,\varepsilon}) &= \frac{1}{N^5}\mathrm{Tr}(D_{j_0}D_{j_6}D_{j_{10}}D_{j_{16}})\mathrm{Tr}(D_{j_4}D_{j_2}D_{j_{14}}D_{j_{12}})\mathrm{Tr}(D_{j_8})\\
&= \frac{1}{N^2}[\tau_N(D_{j_0}D_{j_6}D_{j_{10}}D_{j_{16}})\tau_N(D_{j_4}D_{j_2}D_{j_{14}}D_{j_{12}})\tau_N(D_{j_8})]\\
&\to \frac{1}{N^2}c(j_0,j_6,j_{10},j_{16})c(j_4,j_2,j_{14},j_{12})c(j_8)\\
&\to 0.
\end{aligned}$$

The Gaussian case here is notably easier than the general Wigner case. Pair partitions are quite convenient because they identify pairs of indices, which allows us to order the d_{ij} in cycles as we did above. Then one can show inductively that only non-crossing pairings matter in the limit, as these produce the maximum number of cycles.

For the general, non-Gaussian case, where cumulants of order greater than 2 are allowed, more than two indices may be identified and such cyclic reordering of the d_{ij} is not possible in general. In [**25**], the authors associate a graph to each partition and calculate sharp estimates to prove not only that non-crossing pair partitions are the only non-vanishing contribution, but also stronger forms convergence which depend on more delicate analysis of the contribution of these partitions. Alternatively, one may use concentration of measure [**3**].

- Haar Matrices

The joint distribution of the entries of a Haar-distributed unitary matrix $U_N=(u_{ij})_{i,j\le N}$ is quite complicated. The entries u_{ij} are known to fulfill the Wick formula

$$\mathbb{E}(u_{i_1j_1}\dots u_{i_qj_q}\bar{u}_{i'_1j'_1}\dots\bar{u}_{i'_qj'_q})=\sum_{\rho,\sigma\in S_q}\delta_{i_1i'_{\rho(1)}}\dots\delta_{i_qi'_{\rho(q)}}\delta_{j_1j'_{\sigma(1)}}\dots\delta_{j_qj'_{\sigma(q)}}Wg_N(\rho\sigma^{-1}),$$

where, for each N, the Weingarten function $Wg_N: S_q\to\mathbb{C}$ is some class function whose values depend on characters of representations of symmetric groups.

The leading term of the Weingarten function can be obtained from the asymptotic expansion

$$Wg_N(\sigma)=N^{-n-|\sigma|}\prod_i(-1)^{l(\alpha_i)-1}C_{(l(\alpha_i)-1)}+O(N^{-n-|\sigma|-2}),$$

where the permutation $\sigma = \alpha_1 \dots \alpha_k \in S_n$ is a product of cycles $\alpha_1, \dots, \alpha_k$ of lengths $l(\alpha_i)$, C_k is the k-th Catalan number and $|\sigma|$ is the minimum number of transpositions required to express σ.

One may perform an asymptotic analysis, similar to the ones that we did for the Wigner case, to conclude hat the asymptotic mixed moments of the matrices are computed according to the rules described in Theorem 3.7 (see [**12**], [**14**]).

REMARKS 3.8. *(1). Voiculescu's theorem covers Wishart matrices as well: If p, c are free, c is a circular element and p is a projection, a Wishart matrix of parameter $\lambda > 0$ can essentially be viewed asymptotically as a scalar multiple of cpc^* or pcc^*p (depending on whether $\lambda \leq 1$ or not).*

*(2). There exist a stronger version of freeness, called second-order freeness, which allows to control fluctuations of random matrices. Second order freeness is achieved (see [**3**] or [**25**]) if we slightly strengthen the assumptions on our matrices, by either asking all Wigner random matrices to be Gaussian matrices, or by asking all deterministic matrices to be diagonal. Under such conditions, the empirical eigenvalue distribution of any polynomial*

$$P(X_1^{(N)}, \dots, X_p^{(N)}, U_1^{(N)}, \dots, U_q^{(N)}, D_1^{(N)}, \dots, D_r^{(N)})$$

converges almost surely to the spectral distribution of $P(s_1, \dots, s_p, u_1, \dots u_q, d_1, \dots d_r)$.

Freeness leads quite immediately to matrix-valued freeness: If $A_1, \dots, A_k$ are free in $(\mathcal{A}, \tau)$, then the algebras $M_n(\mathbb{C}) \otimes A_1, \dots, M_n(\mathbb{C}) \otimes A_k$ of matrices with entries in $A_1, \dots, A_k$ are in general not free over $\mathbb{C}$ (w.r.t. $\frac{1}{n}\mathrm{Tr} \otimes \tau$). They are, however $M_n(\mathbb{C})$-free (w.r.t. $id_{M_n(\mathbb{C})} \otimes \tau$). Below is a slightly more general assertion of this simple but fundamental result.

PROPOSITION 3.9. *Let $(\mathcal{A}, \mathbf{F})$ be a $\mathcal{B}$-probability space, and consider the $M_n(\mathcal{B})$-valued probability space $(M_n(\mathbb{C}) \otimes \mathcal{A}, id \otimes \mathbf{F})$. If $A_1, \dots, A_k \subseteq \mathcal{A}$ are $\mathcal{B}$-free, then $(M_n(\mathbb{C}) \otimes A_1), \dots, (M_n(\mathbb{C}) \otimes A_k) \subseteq (M_n(\mathbb{C}) \otimes \mathcal{A})$ are $(M_n(\mathcal{B}))$-free.*

PROOF. Let $a^{(1)}, \dots, a^{(m)} \in M_n(\mathbb{C}) \otimes \mathcal{A}$ be such that $a^{(i)} \in M_n(\mathbb{C}) \otimes A_{j(i)}$ with $j(1) \neq j(2) \neq \dots \neq j(m)$. Observe that

$$\overline{a^{(i)}} = a^{(i)} - (id \otimes \mathbf{F})(a^{(i)}) = ((a_{rs}^{(i)}) - \mathbf{F}(a_{rs}^{(i)}))_{rs \leq n} = (\overline{a_{rs}^{(i)}})_{rs \leq n}.$$

Hence, for all $i, j \leq n$

$$((id \otimes \mathbf{F})((\overline{a^{(1)}}) \cdots (\overline{a^{(m)}})))_{ij} = \sum_{i_1, \dots, i_{m-1}=1}^{n} \mathbf{F}((\overline{a_{ii_1}^{(1)}})(\overline{a_{i_1 i_2}^{(2)}}) \cdots (\overline{a_{i_{m-1}j}^{(m)}}) = 0. \tag{3.11}$$

□

Now that the main aspects of Voiculescu's asymptotic freeness results have been understood, we use Benaych-Georges generalization to rectangular spaces. For our convenience, the version we present is specialized to the situation where the deterministic matrices are blown-up so that condition 3.5 holds trivially. This allows us to directly define free deterministic equivalents.

THEOREM/DEFINITION 3.1 ([**10**]/[**30**]). *Let $N, k \geq 1$ be fixed and let $P_1, \dots, P_k$ be pairwise orthogonal projections such that $I_N = P_1 + \dots + P_k$.*

Let $D_1, \dots, D_r$ be deterministic matrices, with $D_i = P_{h_1(i)} D_i P_{h_2(i)}$ for some $1 \leq h_1(i), h_2(i) \leq k$.

For each $m \geq 1$ and $A \in \{P_1, \dots, P_k, D_1, \dots, D_r\}$, let $A^{(m)} := A \otimes I_m$. Consider now a space $\mathcal{A}_{Nm}$ of $Nm \times Nm$ of random matrices and let $Z_1^{(m)}, \dots, Z_p^{(m)}$ and $U_1^{(m)}, \dots, U_q^{(m)}$ be independent random matrices, such that

(1). For each $i \leq p$, $X_i^{(m)} = P_{j(i)}^{(m)} X_i^{(m)} P_{j(i)}^{(m)}$ is a non-self-adjoint Wigner Matrix when restricted to the compressed space $P_{j(i)}^{(m)} \mathcal{A}_{Nm} P_{j(i)}^{(m)}$ for some $1 \leq j(i) \leq k$.

(2). For each $i \leq q$, $U_i^{(N)} = P_{h(i)}^{(N)} U_i^{(N)} P_{h(i)}^{(N)}$ is a Haar-unitary random matrix when restricted to the compressed space $P_{h(i)} \mathcal{A}_{Nm} P_{h(i)} \otimes \mathcal{M}_m(\mathbb{C})$, for some $1 \leq h(i) \leq k$. Then

$$(X_1^{(m)}, \dots, X_p^{(m)}, U_1^{(m)}, \dots, U_q^{(m)}, D_1^{(m)}, \dots, D_r^{(m)}) \\ \to (c_1, \dots, c_p, u_1, \dots u_q, D_1, \dots D_r)$$

where $c_1, \dots, c_p, u_1, \dots u_q, D_1, \dots D_r$ are elements in a rectangular probability space $(\mathcal{A}, \tau)$ (with $(D_1, \dots, D_r, P_1, \dots, P_k) \subset \mathcal{M}_N(\mathbb{C}) \subset \mathcal{A}$) and such that:

(i). For all $i \leq p$, $c_i = P_{j(i)} c_i P_{j(i)}$ is circular in the compressed space $P_{j(i)} \mathcal{A} P_{j(i)}$.

(ii). For all $i \leq q$, $u_i = p_{h(i)} u_i p_{h(i)}$ is a Haar-unitary in the compressed space $P_{h(i)} \mathcal{A}(\mathbb{C}) P_{h(i)}$.

(iii). The algebras $\langle c_1 \rangle, \dots, \langle c_p \rangle, \langle u_1, u_1^* \rangle, \dots, \langle u_q, u_q^* \rangle, \langle D_1, D_1^* \dots, D_r, D_r^* \rangle$ are free with amalgamation over $\langle P_1, \dots P_k \rangle$.

In particular, for any fixed non-commutative polynomial

$$Q(z_1, z_1^* \dots, z_p, z_p^*, y_1, y_1^*, \dots, y_q, y_q^*, w_1, \dots, w_r),$$

such that the ensemble

$$Q_m := Q(Z_1^{(m)}, (Z_1^{(m)})^* \dots, (Z_p^{(m)})^*, U_1^{(m)}, (U_1^{(m)})^* \dots, (U_q^{(m)})^*, D_1^{(m)}, \dots, D_r^{(m)})$$

is self-adjoint (i.e. $Q_m = Q_m^*$), then Q_m converges in AED to its Free Deterministic Equivalent

$$Q_\infty := Q(c_1, c_1^* \dots, c_p, c_p^*, u_1, u_1^*, \dots, u_q, u_q^*, D_1, \dots, D_r)$$

In order to perform numerical computations, it will be convenient to supress unitaries whenever we deal with randomly rotated deterministic matrices.

3.2. Rotated families of deterministic matrices and elimination of unitaries. If sub-algebras $\langle A, B \rangle \subset \mathcal{A}$ are free from $\langle u, u^* \rangle$, where u is a Haar unitary , and $(a_1, \dots, a_p) \in A^p$, $(b_1, \dots, b_q) \in B^q$, then the joint distribution of $(\tilde{a}_1, \dots, \tilde{a}_p, b_1, \dots b_q)$, where $\tilde{a}_i = u a_i u^*$ is completely determined: $\{\tilde{a}_1, \dots, \tilde{a}_p\}$ and $\{b_1, \dots b_q\}$ are free and $(\tilde{a}_1, \dots, \tilde{a}_p) \sim (a_1, \dots, a_p)$. Roughly speaking, conjugating a family of variables by a free Haar-unitary does not alter the distribution of the family and makes it free from a second family of variables.

The fact that $(\tilde{a}_1, \dots, \tilde{a}_p) \sim (a_1, \dots, a_p)$ is trivial by the tracial property of τ. Hence we only need to show that $\langle u a_1 u^*, \dots, u a_p u^* \rangle$ and $\langle b_1, \dots, b_q \rangle$ are free.

Let $a^{(1)}, \dots, a^{(k)} \in \langle a_1, \dots, a_p \rangle$ and $b^{(1)}, \dots, b^{(k)} \in \langle b_1, \dots, b_q \rangle$. We note that $a^{(j)} \in \langle a_1, \dots, a_p \rangle$ iff $u a^{(j)} u^* \in \langle u a_1 u^*, \dots, u a_p u^* \rangle$. Since $\tau(u a^{(j)} u^*) = \tau(a^{(j)})$ and $\tau(u) = 0 = \tau(u^*)$, we have that $\overline{u a u^*} = \overline{u} \overline{a^{(j)}} \overline{u^*}$.

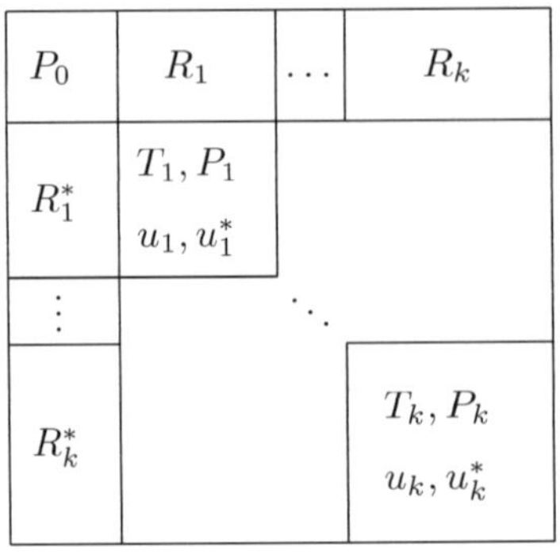

FIGURE 3. Embedding of the operators forming $Q = Q_\infty$ on a Rectangular Space

Hence

$$\begin{aligned}
\tau((\overline{ua^{(1)}u^*})\overline{b^{(1)}}\cdots(\overline{ua^{(k)}u^*})\overline{b^{(k)}}) &= \tau(\overline{u}\,\overline{a^{(1)}}\,\overline{u^*}\,\overline{b^{(1)}}\cdots\overline{u}\,\overline{a^{(k)}}\,\overline{u^*}\,\overline{b^{(k)}}),\\
\tau(\overline{b^{(1)}}(\overline{ua^{(1)}u^*})\cdots\overline{b^{(k)}}(\overline{ua^{(k)}u^*})) &= \tau(\overline{b^{(1)}}\,\overline{u}\,\overline{a^{(1)}}\,\overline{u^*}\cdots\overline{u}\,\overline{a^{(k-1)}}\,\overline{u^*}\,\overline{b^{(k)}}),\\
\tau((\overline{ua^{(1)}u^*})\overline{b^{(1)}}\cdots\overline{b^{(k-1)}}(\overline{ua^{(k)}u^*})) &= \tau(\overline{u}\,\overline{a^{(1)}}\,\overline{u^*}\,\overline{b^{(1)}}\cdots\overline{b^{(k-1)}}\,\overline{u}\,\overline{a^{(k)}}\,\overline{u^*}),\\
\tau(\overline{b^{(1)}}(\overline{ua^{(1)}u^*})\cdots(\overline{ua^{(k-1)}u^*})\overline{b^{(k)}}) &= \tau(\overline{b^{(1)}}\,\overline{u}\,\overline{a^{(1)}}\,\overline{u^*}\cdots\overline{u}\,\overline{a^{(k-1)}}\,\overline{u^*}\,\overline{b^{(k)}}).
\end{aligned}$$

By freeness of $\langle a_1, \dots, a_p, b_1, \dots, b_q\rangle$ and $\{u, u^*\}$ all the RHS expressions vanish and the freeness of $\langle ua_1u^*, \dots, ua_pu^*\rangle$ and $\langle b_1, \dots, b_q\rangle$ is established.

The same statement (with the same proof) holds for several randomly rotated collections:

PROPOSITION 3.10. *Let $A_0, \dots, A_k \subseteq \mathcal{A}$ be $*$-sub-algebras of a $*$-probability space $(\mathcal{A}, \tau)$ and let $u_1, \dots, u_k \in \mathcal{A}$ be Haar-unitary elements, such that $\langle A_0, \dots, A_k\rangle$, $\langle u_1, u_1^*\rangle, \dots, \langle u_k, u_k^*\rangle$ are free. For $0 \le j \le k$, let $(a_1^{(j)}, a_2^{(j)}, \dots, a_{p(j)}^{(j)}) \in A_j^{p(j)}$. Then*

$$(a_1^{(0)}, \dots, a_{p(0)}^{(0)}, u_1a_1^{(1)}u_1^*, \dots, u_1a_{p(1)}^{(1)}u_1^*, \dots, u_ka_{p(k)}^{(k)}u_k^*) \tag{3.12}$$

$$\sim (a_1^{(0)}, \dots, a_{p(0)}^{(0)}, \tilde{a}_1^{(1)}, \dots, \tilde{a}_{p(1)}^{(1)}, \dots, \tilde{a}_{p(k)}^{(k)}), \tag{3.13}$$

where $\langle a_1^{(0)}, \dots, a_{p(0)}^{(0)}\rangle, \langle \tilde{a}_1^{(1)}, \dots, \tilde{a}_{p(1)}^{(1)}\rangle, \dots, \langle \tilde{a}_1^{(k)}, \dots \tilde{a}_{p(k)}^{(k)}\rangle$ are free and $(\tilde{a}_1^{(j)}, \dots \tilde{a}_{p(j)}^{(j)}) \sim (a_1^{(j)}, \dots a_{p(j)}^{(j)})$, for $j \le k$.

By the previous proposition we may, for example, reduce the problem of investigating the distribution of $a + ubu^*$ to that of the sum of free copies $\tilde{a} + \tilde{b}$.

For our example $Q_\infty = \sum_{i=1}^k R_iu_iT_iu_i^*R_i^*$, it is convenient to think that the operators forming Q_∞ (as well as the matrices forming Q_m), are embedded in a rectangular probability space $\mathcal{A}$, as illustrated in Fig. 3 (where $\tilde{N} := N + n_1 + \cdots + n_k$ and $P_0, \dots, P_k \in M_{\tilde{N}} \subset \mathcal{A}$ are such that $Tr(P_0) = N$ and $Tr(P_i) = n_i$). Let $R = \sum_{i=1}^k R_i$, $U = \sum_{i=1}^k u_i$ and $T = \sum_{i=1}^k T_i$, then we have that the desired distribution (of Q_∞) is just that of the element $RUTU^*R^*$ in the compressed space $P_0\mathcal{A}P_0$. Just as in the scalar case, where the unitaries conjugating a variable may be removed, provided that the conjugated variable becomes free from the rest, the element $RUTU^*R^*$ can be thought as $\tilde{R}\tilde{T}\tilde{R}$, where $\tilde{R}, \tilde{T}$ are free over $\langle P_0, \dots, P_k\rangle$.

PROPOSITION 3.11. *Let* $(\mathcal{A}, \mathbf{F})$ *be a* $\mathcal{D} := \langle p_1, \ldots, p_k \rangle$*rectangular probability space. Let* $A_1, A_2 \subset \mathcal{A}$ *and* $U = u_1 + u_2 + \cdots + u_k \in \mathcal{A}$ *be such that* $\langle D_1, D_2 \rangle, U$ *are* $\mathcal{D}$*-free and* $u_j = p_j u_j p_j$ *is a Haar unitary in the compressed space* $p_j \mathcal{A} p_j$*. Then* $D_1, U D_2 U^*$ *are* $\mathcal{D}$*-free.*

PROOF. Just replace τ by $\mathbf{F}$ in Prop. 3.10. □

In [**30**] we showed that the Cauchy Transforms of our FDE's satisfy the Equations in [**17**]. In order to re-derive their equations, we needed to understand how the mixed moments of free random variables are calculated. In the next section we list the main tools for showing this correspondence.

3.3. Combinatorics and cumulants. Recall Definition 2.1. A partition $\pi \in \mathcal{P}(n)$ is *non-crossing* if there is no quadruple of elements $1 \leq i < j < k < l \leq n$ such that $i \sim_\pi k$, $j \sim_\pi l$ and $i \not\sim_\pi j$. The non-crossing partitions of order n form a sub-poset of $\mathcal{P}(n)$ which we denote by $\mathcal{NC}(n)$.

For $n \in \mathbb{N}$, a $\mathbb{C}$-multi-linear map $f : \mathcal{A}^n \to \mathcal{B}$ is called $\mathcal{B}$*-balanced* if it satisfies the $\mathcal{B}$-bilinearity conditions, that for all $b, b' \in \mathcal{B}$, $a_1, \ldots, a_n \in \mathcal{A}$, and for all $r = 1, \ldots, n-1$

$$\begin{aligned} f(ba_1, \ldots, a_n b') &= bf(a_1, \ldots, a_n) b' \\ f(a_1, \ldots, a_r b, a_{r+1}, \ldots, a_n) &= f(a_1, \ldots, a_r, ba_{r+1} \ldots, a_n) \end{aligned}$$

A collection of $\mathcal{B}$-balanced maps $(f_\pi)_{\pi \in \mathcal{NC}}$ is said to be multiplicative with respect to the lattice of non-crossing partitions if, for every $\pi \in \mathcal{NC}$, f_π is computed using the block structure of π in the following way:

- If $\pi = \hat{1}_n \in \mathcal{NC}(n)$, we just write $f_n := f_\pi$.
- If $\hat{1}_n \neq \pi = \{V_1, \ldots, V_k\} \in \mathcal{NC}(n)$, then by a known characterization of $\mathcal{NC}$, there exists a block $V_r = \{s+1, \ldots, s+l\}$ containing consecutive elements. For any such a block we must have

$$f_\pi(a_1, \ldots, a_n) = f_{\pi \backslash V_r}(a_1, \ldots, a_s f_l(a_{s+1}, \ldots, a_{s+l}), a_{s+l+1}, \ldots, a_n),$$

  where $\pi \backslash V_r \in \mathcal{NC}(n-l)$ is the partition obtained from removing the block V_r.

The operator-valued free cumulants $\left(R_\pi^{\mathcal{B}}\right)_{\pi \in \mathcal{NC}}$ are defined as the unique multiplicative family of $\mathcal{B}$-balanced maps satisfying the (operator-valued) moment-cumulant formulas

$$\mathbb{E}(a_1 \ldots a_n) = \sum_{\pi \in \mathcal{NC}(n)} R_\pi^{\mathcal{B}}(a_1, \ldots, a_n)$$

By the cumulants of a tuple $(a_1, \ldots, a_k) \in \mathcal{A}^k$, we mean the collection of all cumulant maps

$$\begin{aligned} R_{i_1, \ldots, i_n}^{\mathcal{B}; a_1, \ldots, a_k} : \quad & \mathcal{B}^{n-1} && \to && \mathcal{B}, \\ & (b_1, \ldots, b_{n-1}) && \mapsto && R_n^{\mathcal{B}}\left(a_{i_1}, b_1 a_{i_2}, \ldots, b_{i_{n-1}} a_{i_n}\right) \end{aligned}$$

for $n \in \mathbb{N}$, $1 \leq i_1, \ldots, i_n \leq k$.

A cumulant map $R_{i_1, \ldots, i_n}^{\mathcal{B}; a_1, \ldots, a_k}$ is mixed if there exists $r < n$ such that $i_r \neq i_{r+1}$. The main feature of the operator-valued cumulants is that they characterize freeness with amalgamation:

THEOREM 3.12 ([**28**]). *The random variables* $a_1, \ldots, a_n$ *are* $\mathcal{B}$*-free iff all their mixed cumulants vanish.*

Other important combinatorial tools (which in particular are used to re-derive the formulas for the DE's from FDE's) are the formulas for computing cumulants of products (see [**23**], [**29**]) and the characterizations of freeness at different levels ([**26**]) in terms of cumulants.

3.4. The analytic subordination phenomena. Like in the scalar case, there are analytical tools to compute operator-valued free convolutions, which are based on the $\mathcal{B}$-valued Cauchy-transform

$$G_x^{\mathcal{B}}(b) = \mathbf{F}((b-x)^{-1}),$$

which maps the operatorial upper half-plane

$$\mathbb{H}^+(\mathcal{B}) := \{b \in \mathcal{B} : \exists \varepsilon > 0 \text{ such that } -i(b-b^*) \geq \varepsilon \cdot 1\}$$

into the lower half-plane $\mathbb{H}^-(\mathcal{B}) = -\mathbb{H}^+(\mathcal{B})$. In the usual settings coming from random matrix models (as we have seen above), our probability space $\mathcal{A}$ may have several operator-valued structures $\mathbf{F}_i : \mathcal{A} \to \mathcal{B}_i$ simultaneously, with $\mathbb{C} = \mathcal{B}_1 \subset \mathcal{B}_2 \subset \cdots \subset \mathcal{B}_k$, and $\mathbf{F}_i \circ \mathbf{F}_{i+1} = \mathbf{F}_i$. We are usually interested ultimately in the scalar-valued distribution, which can be obtained (via Stieltjes inversion) from the Cauchy-transform. The later in turn can be obtained from any "upper" $\mathcal{B}_i$-valued Cauchy transform, as we have that, for all $b \in \mathcal{B}_i$

$$\mathbf{F}_i(G_x^{\mathcal{B}_{i+1}}(b)) = \mathbf{F}_i \circ \mathbf{F}_{i+1}((b-x)^{-1}) = \mathbf{F}_i((b-x)^{-1}) = G_x^{\mathcal{B}_i}(b).$$

A drawback of the operator-valued setting is that, unless we ask $\mathcal{B}$ to be commutative, one can hardly compute explicit distributions: although $\mathcal{B}$-valued generalizations of the R and S-transforms exist ([**33**], [**19**]), the task of explicitly inverting these operator-valued analytic maps is nearly impossible for any non-trivial situation (even for finite dimensional, relatively simple sub-algebras, like $\mathcal{B} = M_2(\mathbb{C})$).

In terms of moments, the operator-valued Cauchy transform is given by

$$G_x^{\mathcal{B}}(b) = \mathbf{F}((b-x)^{-1}) = \sum_{n \geq 0} \mathbf{F}(b^{-1}(xb^{-1})^n)$$

The operator-valued $\mathcal{R}$-transform is defined by

$$\mathcal{R}_x^{\mathcal{B}}(b) = \sum_{n \geq 1} R_n^{\mathcal{B}}(x, bx, \dots, bx).$$

The vanishing of mixed cumulants for free variables implies the additivity of the cumulants, and thus also the additivity of the $\mathcal{R}$-transforms [**33**]: If a_1 and a_2 are $\mathcal{B}$-free then we have for $b \in \mathcal{B}$ that $\mathcal{R}_{a_1+a_2}(b) = \mathcal{R}_{a_1}(b) + \mathcal{R}_{a_2}(b)$.

These transforms satisfy the functional equation

$$G_a^{\mathcal{B}}(b) = \left(\mathcal{R}_a^{\mathcal{B}}(G_a^{\mathcal{B}}(b)) - b\right)^{-1}, \tag{3.14}$$

which was crucial in [**30**] to derive the Equations 1.2 and hence show the correspondence between DE's and FDE's.

Rather than using directly the $\mathcal{R}$-transform, a very powerful method to obtain $\mathcal{B}$-valued free convolutions is based on the analytic subordination phenomena observed by Biane ([**11**], see also [**35**]). In particular, the approach of [**5**] to obtain the subordination functions by iterating analytic maps can be very efficiently performed in the $\mathcal{B}$-valued context.

THEOREM 3.13. **[6]** *Let $(\mathcal{A}, \mathbf{F})$ be a $\mathcal{B}$-valued C^*-probability space and let $x, y \in \mathcal{A}$ be self-adjoint, $\mathcal{B}$-free. There exist an analytic map $\omega : \mathbb{H}^+(\mathcal{B}) \to \mathbb{H}^+(\mathcal{B})$ such that $G_x(\omega(b)) = G_{x+y}(b)$. Furthermore, for any $b \in \mathbb{H}^+(\mathcal{B})$ the subordination function $\omega(b)$ satisfies*

$$\omega(b) = \lim_{n\to\infty} f_b^{\circ n}(w),$$

where, for any $b, w \in \mathbb{H}^+(\mathcal{B})$, $f_b(w) = h_y(h_x(w) + b) + b$ and h is the auxiliary analytic self-map $h_x(b) = (E((b - x)^{-1}))^{-1} - b$ on $\mathbb{H}^+(\mathcal{B})$.

Numerically speaking, going from h_x to G_x and vice-versa is a simple operation. This means that one only needs the individual $\mathcal{B}$-valued Cauchy transforms of x, y (or good approximations of these) to obtain the $\mathcal{B}$-valued Cauchy transform of $x + y$, and hence, its probability distribution. The operator-valued multiplicative convolution can also be numerically approximated (see **[8]**).

In order to implement our main algorithm, we will be interested in the situation described in Prop. 3.9 where $(\mathcal{A}, \mathbf{F})$ is a rectangular probability space and hence our main space $(M_n(\mathbb{C})\otimes\mathcal{A}, id\otimes\mathbf{F})$ consists of $n\times n$ matrices with entries in $\mathcal{A}$, endowed with the entry-wise evaluation of $\mathbf{F} : \mathcal{A} \to \mathcal{B}$. In Section 4 we will use Anderson's self-adjoint linearization trick to obtain the distribution of a polynomial on $\mathcal{B}$-free variables (such as $\tilde{R}$ and $\tilde{T}$, which form Q_∞) from the $(M_n(\mathbb{C}) \otimes \mathcal{B})$-distribution of a specially constructed operator, which depends linearly on the inputs of the polynomial.

In the next section we show how to obtain the $M_n(\mathcal{B})$-valued Cauchy-transforms of such linear elements.

3.5. Linear elements. In a scalar-valued non-commutative probability space $(\mathcal{A}, \tau)$, we have the integral representation of the Cauchy-transform:

$$G_x(z) = \tau((z - x)^{-1}) = \int_{\mathbb{R}} (z - t)^{-1} \mathrm{d}\mu_x(t).$$

Analogously, for linear, self-adjoint elements $D \otimes x$ in a $M_n(\mathbb{C})$-valued probability space $(M_n(\mathbb{C}) \otimes (\mathcal{A}), id_m \otimes \tau)$, we have:

$$G_{D\otimes x}(b) = (id_m \otimes \tau)((b - D \otimes x)^{-1}) = \int_{\mathbb{R}} (b - D \otimes t)^{-1} \mathrm{d}\mu_x(t).$$

The previous integrals can be approximated, for example, by using matrix-valued Riemann sums. In particular, we are able to approximate the $M_n(\mathbb{C})$-valued Cauchy transform of any self-adjoint matrix which depends linearly on a semicircular element s (and hence also for a circular element c which can be viewed as $s_1 + is_2$, for free semi-circulars s_1, s_2). The same can be done if we start with a rectangular probability space.

Let $(\mathcal{A}, \mathbf{F})$ be a $\langle p_1, \dots p_k\rangle$-rectangular probability space and consider the $\mathcal{B}$-valued probability space $(M_m(\mathbb{C})\otimes\mathcal{A}, \mathbf{F}_2)$, where $\mathbf{F}_2 = id_m\otimes\mathbf{F}$ and $\mathcal{B} = (M_m(\mathbb{C})\otimes \langle p_1, \dots p_k\rangle)$

Consider $x \in \mathcal{A}$ of the form $x = \alpha_1 p_1 s_1 p_1 + \dots + \alpha_k p_k s_k p_k$, where $s_i = p_i s_i p_i$ is a semicircular element when restricted to $\mathcal{A}^{(i)}$. Let $D \in M_m(\mathbb{C})$, and let

$b = (b_{ij})_{i,j\leq m} \in \mathcal{B}$, with $b_{ij} = \beta_1^{ij} p_1 + \cdots + \beta_k^{ij} p_k$. Then we have

$$\begin{aligned} G^{\mathcal{B}}_{D\otimes x}(b) &= \mathbf{F}_2((b - D\otimes x)^{-1}) \\ &= (id_m \otimes \mathbf{F})((b - D\otimes x)^{-1}) \\ &= \frac{1}{2\pi}\int_{-2}^{2} [((\beta_1^{ij} - D_{ij}\alpha_1 t)p_1 + \cdots + (\beta_k^{ij} - D_{ij}\alpha_k t)p_k)_{ij}]^{-1}\sqrt{4-t^2}dt. \end{aligned}$$

The case of deterministic matrices is simpler. If we assume that $M_n(\mathbb{C}) \subset \mathcal{A}$ and consider $D = D^* \in M_m(\mathbb{C}) \otimes M_n(\mathbb{C})$. Then $G^{\mathcal{B}}_D(b) = G^{\mathcal{B}}_D(b \otimes I_n)$ is just the partial trace $(id_m \otimes \mathbf{F}_2)((b\otimes I_n - D)^{-1})$.

One should be able to provide a similar trick to approximate Cauchy transforms for elements of the form $D\otimes u + D^* \otimes u^*$. For the moment, we find a way around this problem by removing Haar unitaries, as discussed in Section 3.2.

4. The linearization trick and the main algorithm

One of the main ingredients of our algorithm was already suggested by Voiculescu in his earlier papers on operator-valued free probability: the possibility to transfer questions about the distribution of a polynomial in non-commutative random variables to a question about the matrix-valued distribution of a related polynomial with matrix-valued coefficients, with the advantage of being linear on the non-commutative variables.

The idea was formalized and put into practice by Haagerup and Thorbjørnsen [**21**]. Anderson [**2**] found linearizations which preserve self-adjointness properties. Together with new analytic subordination techniques to compute numerically operator-valued transforms, these self-adjoint linearizations were used in [**6**], to produce a general algorithm to compute asymptotic distributions of non-commutative polynomials evaluated in asymptotically free random matrices.

Our goal in this section is to explain the main ideas behind the algorithm and to show that we can relax some of the hipothesis, so that the resulting algorithm will now cover most of the models in [**17**]. These generalizations are obtained almost literally by replacing the conditional expectation $id\otimes\tau$, used in [**6**],, by some more general $id\otimes \mathbf{F}$ at the very end of the process. For the reader's convenience, we discuss all this in detail for the case of a monomial.

From a linear algebraic point of view, the linearization trick is quite simple and can be traced back to Schuetzenberger in the 40's. It is inspired on the so-called Schur complements which allow to compute matrix inverses by blocks.

If a matrix $M \in M_{n+m}(\mathbb{C})$ has block structure

$$M = \begin{bmatrix} A & B \\ C & D \end{bmatrix},$$

with $D \in M_m(\mathbb{C})$ invertible, then

$$M = \begin{bmatrix} I_n & BD^{-1} \\ 0 & I_m \end{bmatrix} \begin{bmatrix} A - BD^{-1}C & 0 \\ 0 & D \end{bmatrix}, \begin{bmatrix} I_n & 0 \\ D^{-1}C & I_m \end{bmatrix},$$

and hence M is invertible iff the Schur complement $E := A - BD^{-1}C$ is invertible. In such case the inverse has the block structure

$$\begin{bmatrix} A & B \\ C & D \end{bmatrix}^{-1} = \begin{bmatrix} I_n & 0 \\ -D^{-1}C & I_m \end{bmatrix} \begin{bmatrix} E^{-1} & 0 \\ 0 & D^{-1} \end{bmatrix} \begin{bmatrix} I_n & -BD^{-1} \\ 0 & I_m \end{bmatrix} \tag{4.1}$$

$$= \begin{bmatrix} E^{-1} & -E^{-1}BD^{-1} \\ -D^{-1}CE^{-1} & D^{-1} + D^{-1}CE^{-1}BD^{-1} \end{bmatrix} \tag{4.2}$$

The linearization trick exploits the fact that the inverse of the Schur complement E appears alone in the upper corner. We may then choose convenient blocks A, B, C, D so that the Schur complement becomes the resolvent of the non-commutative random variable a that we want to study. Let us first suppose that $a = a_0 \cdots, a_{n+1}$ is a monomial, with $n \geq 0$.

Consider the matrix with non-zero entries

$$M = \begin{bmatrix} & & & & & -a_0 \\ & & & & -a_1 & 1 \\ & & & -a_{n-1} \quad \cdots & 1 & \\ & & -a_n & 1 & & \\ -a_{n+1} & 1 & & & & \end{bmatrix},$$

and let $M = ze_{11} - M'$, where $e_{11} \in M_{n+2}(\mathbb{C}) \otimes \langle 1_{\mathcal{A}} \rangle$ is the upper-left matrix unit.

It is easy to check that the lower right $(n+1) \times (n+1)$ minor D is invertible with inverse given by the lower-triangular matrix

$$D^{-1} = \begin{bmatrix} & & & & & & 1 \\ & & & & & 1 & a_n \\ & & & & 1 & a_{n-1} & a_{n-1}a_n \\ & & & 1 & a_{n-2} & a_{n-2}a_{n-1} & a_{n-2}a_{n-1}a_n \\ & & \cdots & \cdots & & \vdots & \vdots \\ & 1 & a_2 & \cdots & & & \\ 1 & a_1 & a_1a_2 & \cdots & & & a_1a_2\cdots a_n \end{bmatrix}.$$

In fact, this last expression holds even for $D \in M_{n+1}(\mathbb{C}) \otimes \mathcal{A}$, for any unital algebra $\mathcal{A}$ and $a_1, \ldots, a_n \in \mathcal{A}$.

By computing the inverse of $M = ze_{11} - M'$ with Schur complements, the upper-left entry of M^{-1} is simply

$$(A - BD^{-1}C)^{-1} = (z - (0, 0, \ldots, 0, a_0)D^{-1}(0, 0, \ldots, 0, a_0)^{\dagger})^{-1} \tag{4.3}$$

$$= (z - a_0a_1a_2 \cdots a_{n+1})^{-1}. \tag{4.4}$$

In other words, if we let $p := p(a_0, \ldots, a_{n+1}) = a_0a_1 \cdots a_{n+1}$ and we consider the entrywise evaluation of the expectation $id \otimes \tau : M_{n+2}(\mathbb{C}) \otimes \mathcal{A} \to M_{n+2}(\mathbb{C}) \otimes \langle 1_{\mathcal{A}} \rangle$, we get an identity relating $M_{n+2}(\mathbb{C}) \otimes \langle 1_{\mathcal{A}} \rangle$ and $\mathbb{C}$-valued Cauchy transforms:

$$[G^{\mathcal{B}}_{M'}(ze_{11})]_{11} = [(id \otimes \tau)((ze_{11} - M')^{-1})]_{11} = [(id \otimes \tau)(M'^{-1})]_{11} = \tau((z-p)^{-1}) = G^{\mathbb{C}}_p(z).$$

Hence, we have expressed the Cauchy transform of a monomial $p = a_0a_1 \cdots a_{n+1}$ in terms of the $\mathcal{B}$-valued Cauchy transform of an operator which is linear on the a_i's.

We note from the form of M that if p is a self-adjoint monomial (i.e. $a_0a_1 \cdots a_{n+1} = (a_0a_1 \cdots a_{n+1})^*$ then the corresponding linearization M' evcaluated in $a_0, a_1, \ldots, a_{n+1}$ is a self-adjoint, linear element in $M_{n+2}(\mathbb{C}) \otimes \mathcal{A}$. If $a_0, a_1, \ldots, a_{n+1}$ are free, the $M_{n+2}(\mathbb{C})$-Cauchy transform may be computed by the means described in Sections 3.4 and 3.5. This is the algorithm described in [**6**].

In this note we will face the problem that $a_0, a_1, \dots, a_{n+1}$ are not necessarilly $(\mathbb{C}, \tau)$-free, but $(\mathcal{B}, \mathbf{F})$-free. We observe that, in the last equation, we may simply replace $id \otimes \tau$ by $id \otimes \mathbf{F}$. We obtain again a relation between Cauchy transforms. For the case where the expectations come from a rectangular space, we can still proceed by the means of Sections 3.4 and 3.5. Examples will be given at the end of the section.

Now that we know how to linearize monomials, a general polinomial can be linearized in terms of the linearizations of its monomials by using a finite combination of the following rules [**2**]:

If $P = P_1 + \cdots + P_k$ and each P_j has a linearization

$$L_{P_j} = \begin{bmatrix} 0 & u_j \\ v_j & Q_j \end{bmatrix},$$

then a linearization of P is given by

$$L_P = \begin{bmatrix} 0 & u_1 & \cdots & u_k \\ v_1 & & & \\ \vdots & & \ddots & \\ v_k & & & Q_k \end{bmatrix}.$$

Finally, if P is self-adjoint, we may view it as $P = q + q^*$ for $q = P/2$. If

$$L_q = \begin{bmatrix} 0 & u \\ v & Q \end{bmatrix}$$

is a linearization of q then

$$L_P = \begin{bmatrix} 0 & u & v^* \\ u^* & 0 & Q^* \\ v & Q & 0 \end{bmatrix}$$

is a self-adjoint linearization of P.

The algorithm in [**6**] allows to obtain the Cauchy transform of any self-adjoint polynomial $p = p^*$ evaluated on free self-adjoint entries. We relax this situation slightly by asking the polynomial to be self-adjoint only after evaluating it on the specific non-commutative random variables of interest, which may now be non-selfadjoint and we assume to be $\mathcal{B}$-free with respect to some conditional expectation $\mathbf{F}$.

PROPOSITION 4.1. *Let $(\mathcal{A}, \mathcal{B})$ be a $\mathcal{B}$-probability space and let $x_1, \dots, x_n \in \mathcal{A}$. Let $P = P(x_1, \dots, x_n) \in \mathcal{B}\langle x_1, \dots x_n, x_1^*, \dots x_n^* \rangle$ be a self-adjoint $\mathcal{B}$-valued polynomial in $x_1, \dots, x_n$ and their adjoints. There exist $m \geq 1$ and an element $L_P \in M_m(\mathbb{C}) \otimes \mathcal{A}$ such that:*

(1) *$L_P = c_1 \otimes x_1 + c_1^* \otimes x_1^* + \dots c_n \otimes x_n + c_n^* \otimes x_n^* + c \in M_m(\mathbb{C}) \otimes \mathcal{A}$, with $c \in M_m(\mathbb{C}) \otimes \mathcal{B}$ and, for $i \geq 1$ $c_i \in M_m(\mathbb{C})$.*

(2) *If $\Lambda_\varepsilon(b) = diag(b, i\varepsilon, i\varepsilon, \dots, i\varepsilon) \in M_m(\mathbb{C}) \otimes \mathcal{B}$, then*

$$G_P^{\mathcal{B}}(b) = \lim_{\varepsilon \downarrow 0} (G_{L_P}^{M_m(\mathbb{C}) \otimes \mathcal{B}}(\Lambda_\varepsilon(b)))_{11}.$$

PROOF. The general case follows the same lines as the case of a monomial. For the second part, we must consider $\Delta_\varepsilon(b)$ as in [[**6**], Cor. 3.6] so that the argument belongs to the operatorial upper-half-plane, which is the right domain of the Cauchy-transform for a later application of Theorem 3.13). □

REMARK 4.2. *Since we are able to compute operator-valued Cauchy transforms of arbitrary deterministic matrices (as these are just partial traces), the products of deterministic matrices do not really bother us. We should use the linearization trick only to transform the polynomial into a polynomial with matrix coefficients which is linear in the variables which correspond to random matrices but needs not necessarily to be linear on the variables corresponding to deterministic matrices.*

4.1. Examples from wireless communications. Now we consider some matrix models from [**17**]. Understanding these models and their deterministic equivalents was one of the main motivations of our work. After each model, we discuss embeddings of the matrices in rectangular spaces, we then discuss the FDE and we give a linearization which allows to plot the distribution.

4.1.1. *Unitary precoded channels* [**16**]. For the model

$$Q = \sum_{i=1}^{k} R_i U_i T_i U_i^* R_i^*,$$

we already discussed its embedding in a rectangular space and its FDE $Q_N = \tilde{R}\tilde{T}\tilde{R}^*$, so we are only missing its linearization. It is very simple, namely

$$L_{P_\infty} = \begin{bmatrix} 0 & 0 & -\tilde{R} \\ 0 & -\tilde{T} & 1 \\ -\tilde{R}^* & 1 & 0 \end{bmatrix},$$

where each entry is really an $\tilde{N} \times \tilde{N}$ block, with $\tilde{N} = N_0 + n_1 \cdots + n_k$. The individual $M_3(\mathcal{B})$-valued Cauchy transforms of the self-adjoint elements

$$L_1 = \begin{bmatrix} 0 & 0 & -\tilde{R} \\ 0 & 0 & 1 \\ -\tilde{R}^* & 1 & 0 \end{bmatrix}, \quad L_2 = \begin{bmatrix} 0 & 0 & 0 \\ 0 & -\tilde{T} & 0 \\ 0 & 0 & 0 \end{bmatrix},$$

can be computed by performing partial traces, as explained in Section 3.5.

Fig. 2 in the Introduction shows the implementation of our algorithm for this case.

4.1.2. *Correlated MIMO multiple access channels* [**15**]. .

Let us go back to the model

$$P = \sum_{i=1}^{k} R_i Z_i T_i Z_i^* R_i.$$

In order to achieve asymptotic freeness we embed the matrices in a rectangular space exactly as we did in the previous case. The FDE will be then

$$P_\infty = \sum_{i=1}^{k} R_i c_i T_i c_i^* R_i^*,$$

and the linearization yields

$$L_{P_\infty} = \begin{bmatrix} 0 & 0 & 0 & 0 & -R \\ 0 & 0 & 0 & -c & 1 \\ 0 & 0 & -T & 1 & 0 \\ 0 & -c^* & 1 & 0 & 0 \\ -R^* & 1 & 0 & 0 & 0 \end{bmatrix},$$

where each entry is again $\tilde{N} \times \tilde{N}$, with $\tilde{N} = N + n_1 \cdots + n_k$. The individual $M_5(\mathcal{B})$-valued Cauchy-transforms of the self-adjoint elements

$$L_1 = \begin{bmatrix} 0 & 0 & 0 & 0 & -R \\ 0 & 0 & 0 & 0 & 1 \\ 0 & 0 & -T & 1 & 0 \\ 0 & 0 & 1 & 0 & 0 \\ -R^* & 1 & 0 & 0 & 0 \end{bmatrix}, \quad L_2 = \begin{bmatrix} 0 & 0 & 0 & 0 & 0 \\ 0 & 0 & 0 & -c & 0 \\ 0 & 0 & 0 & 0 & 0 \\ 0 & -c^* & 0 & 0 & 0 \\ 0 & 0 & 0 & 0 & 0 \end{bmatrix},$$

can be computed by, respectively, performing a partial trace and approximating by matrix-valued Riemann sums (or, alternatively, by using the method in [**22**]), as explained in Section 3.5.

4.1.3. *Frequency selective MIMO systems* [**18**]. Let $k \geq 1$ be fixed again and consider now the model

$$P_N = \sum_{i=1}^{k} (R_i Z_i T_i) \sum_{j=1}^{k} (T_j^* Z_j^* R_j^*),$$

where the individual matrices are as in the previous example and additionally $n_i = n$ for all $i \leq k$.

We embed again R_i, Z_i in a rectangular space in such a way that $P_i R_i P_0 = R_i$ and $P_i Z_i P_i = Z_i$, but this time we put $P_0 T_i P_i = T_i$.

If (as in the first example), we put again $R = \sum R_i$, $T = \sum T_i$, $c = \sum c_i$, our FDE can be compactly written as

$$P_\infty = RcTT^* c^* R^*.$$

The linearization will be very similar to the one in the previous case, the main difference is the way in which we have embedded the matrices T_i. We get

$$L_{P_\infty} = \begin{bmatrix} 0 & 0 & 0 & 0 & -R \\ 0 & 0 & 0 & -c & 1 \\ 0 & 0 & -TT^* & 1 & 0 \\ 0 & -c^* & 1 & 0 & 0 \\ -R^* & 1 & 0 & 0 & 0 \end{bmatrix},$$

where each entry is $\tilde{N} \times \tilde{N}$, with $\tilde{N} = N + kn$. The individual $M_5(\mathcal{D})$-valued Cauchy-transforms of the self-adjoint elements

$$L_1 = \begin{bmatrix} 0 & 0 & 0 & 0 & -R \\ 0 & 0 & 0 & 0 & 1 \\ 0 & 0 & -TT^* & 1 & 0 \\ 0 & 0 & 1 & 0 & 0 \\ -R^* & 1 & 0 & 0 & 0 \end{bmatrix}, \quad L_2 = \begin{bmatrix} 0 & 0 & 0 & 0 & 0 \\ 0 & 0 & 0 & -c & 0 \\ 0 & 0 & 0 & 0 & 0 \\ 0 & -c^* & 0 & 0 & 0 \\ 0 & 0 & 0 & 0 & 0 \end{bmatrix},$$

can be again computed as explained in Section 3.5.

References

[1] A. Lu, X. Gao, and C. Xiao, A free deterministic equivalent for the capacity of mimo mac with distributed antenna sets, *2015 IEEE International Conference on Communications* (2015) 1751–1756.

[2] G. W. Anderson, *Convergence of the largest singular value of a polynomial in independent Wigner matrices*, Ann. Probab. **41** (2013), no. 3B, 2103–2181. MR3098069

[3] G. W. Anderson, A. Guionnet, and O. Zeitouni, *An introduction to random matrices*, Cambridge Studies in Advanced Mathematics, vol. 118, Cambridge University Press, Cambridge, 2010. MR2760897

[4] O. Arizmendi, I. Nechita, and C. Vargas, *On the asymptotic distribution of block-modified random matrices*, J. Math. Phys. **57** (2016), no. 1, 015216, 25. MR3432744

[5] S. T. Belinschi and H. Bercovici, *A new approach to subordination results in free probability*, J. Anal. Math. **101** (2007), 357–365. MR2346550

[6] S. T. Belinschi, T. Mai, and R. Speicher, *Analytic subordination theory of operator-valued free additive convolution and the solution of a general random matrix problem*, J. Reine Angew. Math. **732** (2017), 21–53. MR3717087

[7] S. Belinschi, P. Sniady, and R. Speicher, Eigenvalues of non-hermitian random matrices and brown measure of non-normal operators: hermitian reduction and linearization method, preprint (2015).

[8] S. Belinschi, R. Speicher, J. Treilhard, and C. Vargas, Operator-valued free multiplicative convolution: analytic subordination theory and applications to random matrix theory, *Internat. Math. Res. Notices* (2014).

[9] F. Benaych-Georges, *Rectangular random matrices, related convolution*, Probab. Theory Related Fields **144** (2009), no. 3-4, 471–515. MR2496440

[10] F. Benaych-Georges, *Rectangular random matrices, entropy, and Fisher's information*, J. Operator Theory **62** (2009), no. 2, 371–419. MR2552088

[11] P. Biane, *Processes with free increments*, Math. Z. **227** (1998), no. 1, 143–174. MR1605393

[12] B. Collins, *Moments and cumulants of polynomial random variables on unitary groups, the Itzykson-Zuber integral, and free probability*, Int. Math. Res. Not. **17** (2003), 953–982. MR1959915

[13] B. Collins, *Moments and cumulants of polynomial random variables on unitary groups, the Itzykson-Zuber integral, and free probability*, Int. Math. Res. Not. **17** (2003), 953–982. MR1959915

[14] B. Collins and P. Śniady, *Integration with respect to the Haar measure on unitary, orthogonal and symplectic group*, Comm. Math. Phys. **264** (2006), no. 3, 773–795. MR2217291

[15] R. Couillet, M. Debbah, and J. W. Silverstein, *A deterministic equivalent for the analysis of correlated MIMO multiple access channels*, IEEE Trans. Inform. Theory **57** (2011), no. 6, 3493–3514. MR2817033

[16] R. Couillet, J. Hoydis, and M. Debbah, A deterministic equivalent approach to the performance analysis of isometric random precoded systems, *IEEE Trans. Inf. Theory* (2011).

[17] R. Couillet and M. Debbah, *Random matrix methods for wireless communications*, Cambridge University Press, Cambridge, 2011. MR2884783

[18] F. Dupuy and P. Loubaton, Mutual information of frequency selective MIMO systems: an asymptotic approach, *Ann. Appl. Prob.* **17** (2007), 875–930.

[19] K. J. Dykema, *On the S-transform over a Banach algebra*, J. Funct. Anal. **231** (2006), no. 1, 90–110. MR2190164

[20] V. L. Girko, *Theory of stochastic canonical equations. Vol. I*, Mathematics and its Applications, vol. 535, Kluwer Academic Publishers, Dordrecht, 2001. MR1887675

[21] U. Haagerup and S. Thorbjørnsen, *A new application of random matrices:* $\mathrm{Ext}(C^*_{\mathrm{red}}(F_2))$ *is not a group*, Ann. of Math. (2) **162** (2005), no. 2, 711–775. MR2183281

[22] J. W. Helton, R. Rashidi Far, and R. Speicher, *Operator-valued semicircular elements: solving a quadratic matrix equation with positivity constraints*, Int. Math. Res. Not. IMRN **22** (2007), Art. ID rnm086, 15. MR2376207

[23] B. Krawczyk and R. Speicher, *Combinatorics of free cumulants*, J. Combin. Theory Ser. A **90** (2000), no. 2, 267–292. MR1757277

[24] V. Marchenko and L. Pastur. Distribution of eigenvalues for some sets of random matrices, *Math. USSR-Sbornik* **1**, (1967), 457–483.

[25] J. A. Mingo and R. Speicher, *Sharp bounds for sums associated to graphs of matrices*, J. Funct. Anal. **262** (2012), no. 5, 2272–2288. MR2876405

[26] A. Nica, D. Shlyakhtenko, and R. Speicher, *Operator-valued distributions. I. Characterizations of freeness*, Int. Math. Res. Not. **29** (2002), 1509–1538. MR1907203

[27] D. Shlyakhtenko, *Random Gaussian band matrices and freeness with amalgamation*, Internat. Math. Res. Notices **20** (1996), 1013–1025. MR1422374

[28] R. Speicher, *Combinatorial theory of the free product with amalgamation and operator-valued free probability theory*, Mem. Amer. Math. Soc. **132** (1998), no. 627, x+88. MR1407898

[29] R. Speicher, *A conceptual proof of a basic result in the combinatorial approach to freeness*, Infin. Dimens. Anal. Quantum Probab. Relat. Top. **3** (2000), no. 2, 213–222. MR1812697

[30] R. Speicher and C. Vargas, *Free deterministic equivalents, rectangular random matrix models, and operator-valued free probability theory*, Random Matrices Theory Appl. **1** (2012), no. 2, 1150008, 26. With an appendix by Tobias Mai. MR2934714

[31] D. Voiculescu, *Symmetries of some reduced free product C^*-algebras*, Operator algebras and their connections with topology and ergodic theory (Buşteni, 1983), Lecture Notes in Math., vol. 1132, Springer, Berlin, 1985, pp. 556–588. MR799593

[32] D. Voiculescu, *Limit laws for random matrices and free products*, Invent. Math. **104** (1991), no. 1, 201–220. MR1094052

[33] D. Voiculescu, *Operations on certain non-commutative operator-valued random variables*, Astérisque **232** (1995), 243–275. Recent advances in operator algebras (Orléans, 1992). MR1372537

[34] D. Voiculescu, *The coalgebra of the free difference quotient and free probability*, Internat. Math. Res. Notices **2** (2000), 79–106. MR1744647

[35] D. Voiculescu, *Analytic subordination consequences of free Markovianity*, Indiana Univ. Math. J. **51** (2002), no. 5, 1161–1166. MR1947871

[36] E. P. Wigner, *On the distribution of the roots of certain symmetric matrices*, Ann. of Math. (2) **67** (1958), 325–327. MR0095527

CONACYT-CIMAT, Centro de Investigación en Matemáticas A. C., Guanajuato, México

Email address: carlosv@cimat.mx

Selected Published Titles in This Series

709 **Fernando Galaz-García, Juan Carlos Pardo Millán, and Pedro Solórzano, Editors,** Contributions of Mexican Mathematicians Abroad in Pure and Applied Mathematics, 2018

706 **Yeonhyang Kim, Sivaram K. Narayan, Gabriel Picioroaga, and Eric S. Weber, Editors,** Frames and Harmonic Analysis, 2018

705 **Graham J. Leuschke, Frauke Bleher, Ralf Schiffler, and Dan Zacharia, Editors,** Representations of Algebras, 2018

704 **Alain Escassut, Cristina Perez-Garcia, and Khodr Shamseddine, Editors,** Advances in Ultrametric Analysis, 2018

703 **Andreas Malmendier and Tony Shaska, Editors,** Higher Genus Curves in Mathematical Physics and Arithmetic Geometry, 2018

702 **Mark Grant, Gregory Lupton, and Lucile Vandembroucq, Editors,** Topological Complexity and Related Topics, 2018

701 **Joan-Carles Lario and V. Kumar Murty, Editors,** Number Theory Related to Modular Curves, 2018

700 **Alexandre Girouard, Dmitry Jakobson, Michael Levitin, Nilima Nigam, Iosif Polterovich, and Frédéric Rochon, Editors,** Geometric and Computational Spectral Theory, 2017

699 **Mark L. Agranovsky, Matania Ben-Artzi, Catherine Bénéteau, Lavi Karp, Dmitry Khavinson, Simeon Reich, David Shoikhet, Gilbert Weinstein, and Lawrence Zalcman, Editors,** Complex Analysis and Dynamical Systems VII, 2017

698 **Alexander M. Blokh, Leonid A. Bunimovich, Paul H. Jung, Lex G. Oversteegen, and Yakov G. Sinai, Editors,** Dynamical Systems, Ergodic Theory, and Probability: in Memory of Kolya Chernov, 2017

697 **Fabrizio Broglia, Françoise Delon, Max Dickmann, Danielle Gondard-Cozette, and Victoria Ann Powers, Editors,** Ordered Algebraic Structures and Related Topics, 2017

696 **Ara S. Basmajian, Yair N. Minsky, and Alan W. Reid, Editors,** In the Tradition of Ahlfors–Bers, VII, 2017

695 **Katrina Barron, Elizabeth Jurisich, Antun Milas, and Kailash Misra, Editors,** Lie Algebras, Vertex Operator Algebras, and Related Topics, 2017

694 **Manjul Bhargava, Robert Guralnick, Gerhard Hiss, Klaus Lux, and Pham Huu Tiep, Editors,** Finite Simple Groups: Thirty Years of the Atlas and Beyond, 2017

693 **Michael Cwikel and Mario Milman, Editors,** Functional Analysis, Harmonic Analysis, and Image Processing, 2017

692 **Anatole Katok, Yakov Pesin, and Federico Rodriguez Hertz, Editors,** Modern Theory of Dynamical Systems, 2017

691 **Farrell Brumley, Maria Paula Gómez Aparicio, and Alberto Mínguez, Editors,** Around Langlands Correspondences, 2017

690 **Andrés Eduardo Caicedo, James Cummings, Peter Koellner, and Paul B. Larson, Editors,** Foundations of Mathematics, 2017

689 **Erica Flapan, Allison Henrich, Aaron Kaestner, and Sam Nelson, Editors,** Knots, Links, Spatial Graphs, and Algebraic Invariants, 2017

688 **Jeffrey Bergen, Stefan Catoiu, and William Chin, Editors,** Groups, Rings, Group Rings, and Hopf Algebras, 2017

687 **Fernanda Botelho, Raena King, and T. S. S. R. K. Rao, Editors,** Problems and Recent Methods in Operator Theory, 2017

686 **Alp Bassa, Alain Couvreur, and David Kohel, Editors,** Arithmetic, Geometry, Cryptography and Coding Theory, 2017

For a complete list of titles in this series, visit the AMS Bookstore at **www.ams.org/bookstore/conmseries/**.